RIVER PUBLISHERS SERIES IN MATHEMATICAL AND ENGINEERING SCIENCES

Series Editors

MANGEY RAM
Graphic Era University, India

TADASHI DOHI
Hiroshima University, Japan

ALIAKBAR MONTAZER HAGHIGHI
Prairie View Texas A&M University, USA

Mathematics is the basis of all disciplines in science and engineering. Especially applied mathematics has become complementary to every branch of engineering sciences. The purpose of this book series is to present novel results in emerging research topics on engineering sciences, as well as to summarize existing research. It engrosses mathematicians, statisticians, scientists and engineers in a comprehensive range of research fields with different objectives and skills, such as differential equations, finite element method, algorithms, discrete mathematics, numerical simulation, machine leaning, probability and statistics, fuzzy theory, etc.

Books published in the series include professional research monographs, edited volumes, conference proceedings, handbooks and textbooks, which provide new insights for researchers, specialists in industry, and graduate students.

Topics covered in the series include, but are not limited to:

- Advanced mechatronics and robotics
- Artificial intelligence
- Automotive systems
- Discrete mathematics and computation
- Fault diagnosis and fault tolerance
- Finite element methods
- Fuzzy and possibility theory
- Industrial automation, process control and networked control systems
- Intelligent control systems
- Neural computing and machine learning
- Operations research and management science
- Optimization and algorithms
- Queueing systems
- Reliability, maintenance and safety for complex systems
- Resilience
- Stochastic modelling and statistical inference
- Supply chain management
- System engineering, control and monitoring
- Tele robotics, human computer interaction, human-robot interaction

For a list of other books in this series, visit www.riverpublishers.com

Special Functions and their Applications

Special Functions and their Applications

Bipin Singh Koranga
Kirori Mal College, Delhi University, India

Sanjay Kumar Padaliya
S.G.R.R. (P.G.) College, India

Vivek Kumar Nautiyal
Babasaheb Bhimrao Ambedkar University, India

LONDON AND NEW YORK

Published 2021 by River Publishers
River Publishers
Alsbjergvej 10, 9260 Gistrup, Denmark
www.riverpublishers.com

Distributed exclusively by Routledge
4 Park Square, Milton Park, Abingdon, Oxon OX14 4RN
605 Third Avenue, New York, NY 10017, USA

Special Functions and their Applications / by Bipin Singh Koranga, Sanjay Kumar Padaliya, Sanjay Kumar Padaliya.

© 2021 River Publishers. All rights reserved. No part of this publication may be reproduced, stored in a retrieval systems, or transmitted in any form or by any means, mechanical, photocopying, recording or otherwise, without prior written permission of the publishers.

Routledge is an imprint of the Taylor & Francis Group, an informa business

ISBN 978-87-7022-626-4 (print)

While every effort is made to provide dependable information, the publisher, authors, and editors cannot be held responsible for any errors or omissions.

Contents

Preface ix

List of Tables xi

1 **The Gamma Function** 1
  1.1 Definition of Gamma Function 1
  1.2 Gamma Function and Some Relations 3
  1.3 The Logarithmic Derivative of the Gamma Function . . 6
  1.4 Asymptotic Representation of the Gamma Function for Large $|z|$ 10
  1.5 Definite Integrals Related to the Gamma Function . . . 11
  1.6 Exercises 12

2 **The Probability Integral and Related Functions** 15
  2.1 The Probability Integral and its Basic Properties 15
  2.2 Asymptotic Representation of Probability Integral for Large $|z|$ 17
  2.3 The Probability Integral of Imaginary Argument 18
  2.4 The Probability Fresnel Integrals 20

vi *Contents*

    2.5   Application to Probability Theory 23

    2.6   Application to the Theory of Heat Conduction 24

    2.7   Application to the Theory of Vibrations 26

    2.8   Exercises 28

3 Spherical Harmonics Theory   **31**

    3.1   Introduction 31

    3.2   The Hypergeometric Equation and its Series Solution . 32

    3.3   Legendre Functions 35

    3.4   Integral Representations of the Legendre Functions . . 37

    3.5   Some Relations Satisfied by the Legendre Functions . . 39

    3.6   Workskian of Pairs of Solutions of Legendre's Equation 40

    3.7   Recurrence Relations for the Legendre Functions . . . 42

    3.8   Associated Legendre Functions 44

    3.9   Exercises 46

4 Bessel Function   **49**

    4.1   Bessel Functions 49

    4.2   Generating Function 54

    4.3   Recurrence Relations 57

    4.4   Orthonormality 59

    4.5   Application to the Optical Fiber 60

    4.6   Exercises 62

5 Hermite Polynomials   **65**

    5.1   Hermite Functions 65

    5.2   Generating Function 69

5.3	Recurrence Relations	70
5.4	Rodrigues Formula	73
5.5	Orthogonality and Normalilty	74
5.6	Application to the Simple Harmonic Oscillator	76
5.7	Exercises	78

6 Laguerre Polynomials — 81

6.1	Laguerre Functions	81
6.2	Generating Function	85
6.3	Recurrence Relations	87
6.4	Rodrigues Formula	91
6.5	Orthonormality	92
6.6	Application to the Hydrogen Atom	94
6.7	Associated Laguerre Polynomials	98
	6.7.1 Properties of Associated Laguerre Polynomials	102
6.8	Exercises	102

Bibliography — 105

Index — 107

About the Authors — 109

Preface

We feel great pleasure in bringing out the first edition of the book "Special Functions and their Applications." This book has been written especially in accordance with the latest and modified syllabus framed for UG and PG students. A reasonably wide coverage in sufficient depth has been attempted. The book contains sufficient number of problems. We hope that if a student goes through all these, he or she would appreciate and enjoy the subject.

The work is dedicated to our past students whose inspiration motivated us to do this work without any stress and strain. We wish to express our indebtedness to numerous authors of those books that were consulates during the preparation of the matter. We feel great pleasure to express deepest sense of gratitude, respect, and honor to Prof. Uma Sankar, IIT Bombay and Prof. Mohan Narayan, Mumbai University for providing valuable guidance and for their patronly behavior. We would like to thank several of our colleagues at Kirori Mal College and SGRR(PG) College, Dehradun. We would like to thank Dr. Vinod Kumar, Assistant Prof. University of Lucknow, Dr. Imran Khan, Assistant Prof. Ramjas College and Dr. Shushil, Assistant Prof. Hindu College, for making useful suggestions during the preparation of this manuscript.

Errors might have crept in here and there in spite of care to avoid them. We will be very grateful for bringing them to our notice. Suggestions or criticisms toward further improvement of the book shall be gratefully acknowledged.

We shall appreciate receiving comments and suggestions, which can be sent to the following emails:
bipiniitb@rediffmail.com
spadaliya12@rediffmail.com
viveknautiyal01@gmail.com

Thanks are due to Prof. Angelo Galanty, University of L'Aquila and Dr. Rakesh Joshi, Australia's Global University for encouragement. We are deeply grateful to our family members who always have been a source of inspiration for us. Last, but not the least, we are thankful to the publisher of this book.

Dr. Bipin Singh Koranga
Dr. Vivek Kumar Nautiyal
Dr. Sanjay Kumar Padaliya

List of Tables

Table 5.1 Some Hermite Polynomials 70

Table 6.1 Some Laguerre Polynomials 88

1

The Gamma Function

1.1 Definition of Gamma Function

One of the simplest and most important special functions is the gamma function. Its properties are used for the study of many other special functions, the cylinder functions, and the hypergeometric function. The gamma function is usually studied in courses on complex variable theory and in advanced calculus. The gamma function is defined by the following:

$$\Gamma(z) = \int_0^\infty e^{-t} t^{z-1} dt, \quad \operatorname{Re} z \succ 0, \tag{1.1}$$

whenever the complex variable z has a positive real part $\operatorname{Re} z$. We can write (1.1) as the sum of two integrals,

$$\Gamma(z) = \int_0^1 e^{-t} t^{z-1} dt + \int_1^\infty e^{-t} t^{z-1} dt, \tag{1.2}$$

where it can easily be shown that the first integral defines a function $P(z)$, which is analytic in the half plane $\operatorname{Re} z > 0$, while the second integral defines an entire function. If follows that the function $\Gamma(z) = P(z) + Q(z)$ is analytic in the half-plane $\operatorname{Re} z > 0$. The values of $\Gamma(z)$

2 The Gamma Function

in the rest of the complex plane can be found by analytic continuation of the function defined by (1.1). First, we replace the exponential in the integral for $P(z)$ by its power expansion and we integrate term by term, we obtaining

$$P(z) = \int_0^1 t^{z-1} dt \sum_{k=0}^{\infty} \frac{(-1)^k}{k!} t^k = \sum_{k=0}^{\infty} \frac{(-1)^k}{k!} \int_0^1 t^{k+z-1} dt$$

$$= \sum_{k=0}^{\infty} \frac{(-1)^k}{k!} \frac{1}{z+k}, \qquad (1.3)$$

where it is permissible to reverse the order of integration and summation since

$$\int_0^1 |t^{z-1}| dt \sum_{k=0}^{\infty} \left| \frac{(-1)^k}{k!} t^k \right| = \int_0^1 t^{x-1} dt \sum_{k=0}^{\infty} \frac{t^k}{k!} = \int_0^1 e^t t^{x-1} dt < \infty$$

The last integral converges for $x = \operatorname{Re} z > 0$. The terms of the series (1.3) are analytic functions of z, if $z \neq 0, -1, -2...$. In the region

$$|z+k| \geqslant \delta > 0, \qquad k = 0, 1, 2....,$$

(1.3) is expressed by the convergent series

$$\sum_{k=0}^{\infty} \frac{1}{k! \delta},$$

and, hence, it is uniformly convergent in this region. We conclude that the sum of the series (1.3) is a meromorphic function with simple poles at the points $z = 0, -1, -2...$. For $\operatorname{Re} z > 0$, this function concides with the integral $P(z)$ and, hence, is the analytic continuation of $P(z)$. The function $\Gamma(z)$ differs from $P(z)$ by the term $Q(z)$, which is an entire function. Therefore, $\Gamma(z)$ is a meromorphic function of the complex variable z, with simple poles at the points $z = 0, -1, -2....$ An analytic expression for $\Gamma(z)$, suitable for defining $\Gamma(z)$ in the whole complex plane, is given by

$$\Gamma(z) = \sum_{k=0}^{\infty} \frac{(-1)^k}{k!} \frac{1}{z+k} + \int_1^{\infty} e^{-t} t^{z-1} dt, \quad z \neq 0, -1, -2, ... \quad (1.4)$$

It follows from (1.4) that $\Gamma(z)$ that has the representation

$$\Gamma(z) = \sum_{k=0}^{\infty} \frac{(-1)^n}{n!} \frac{1}{z+n} + \Omega(z+n) \quad (1.5)$$

in a neighborhood of the pool $z = -n$ ($n = 0, 1, 2...$), with regular part $\Omega(z+n)$.

1.2 Gamma Function and Some Relations

We now consider three basic relations satisfied by the gamma function

$$\Gamma(z+1) = z\Gamma(z), \quad (1.6)$$

$$\Gamma(z)\Gamma(1-z) = \frac{\pi}{sin\pi z}, \quad (1.7)$$

$$2^{2z-1}\Gamma(z)\Gamma(z+\frac{1}{2}) = \sqrt{\pi}\Gamma(2z). \quad (1.8)$$

These expressions play an important role in various transformations and calculations involving $\Gamma(z)$. To prove (1.1), we assume that $\mathrm{Re} > 0$ and use the integral representation (1.1). An integration by parts gives

$$\Gamma(z+1) = \int_0^{\infty} e^{-t} t^z dt = -e^{-t} t^z \big|_0^{\infty} + z \int_0^{\infty} e^{-t} t^{z-1} dt = z\Gamma(z). \quad (1.9)$$

4 The Gamma Function

The validity of this result for arbitrary complex $z \neq 0, -1, -2, \ldots$ is an immediate consequence of the principle of analytic continuation; both sides of the expression are analytic everywhere except at the points $z = 0, -1, -2\ldots$

To derive (1.7), we assume that $0 < \operatorname{Re} z < 1$, and, again, we use (1.6); we get

$$\Gamma(z)\Gamma(1-z) = \int_0^\infty \int_0^\infty e^{-(s+t)} s^{-z} t^{z-1} ds dt. \qquad (1.10)$$

We introduce the new variables

$$u = s+t, \qquad\qquad v = \frac{t}{s}.$$

We find that

$$\Gamma(z)\Gamma(1-z) = \int_0^\infty \int_0^\infty e^{-u} v^{z-1} \frac{du\, dv}{1+v} = \int_0^\infty \frac{v^{z-1}}{1+v} dv = \frac{\pi}{\sin \pi z}.$$

The above expression is valid in the complex plane except at the points $z = 0, \pm 1, \pm 2, \ldots$

To prove (1.8), we assume that $\operatorname{Re} z > 0$ and then use (1.6), and we get

$$2^{2z-1}\Gamma(z)\Gamma(z+1) = \int_0^\infty \int_0^\infty e^{-(s+t)} (2\sqrt{st})^{2z-1} t^{-1/2} ds dt$$

$$= 4 \int_0^\infty \int_0^\infty e^{-(\alpha^2 + \beta^2)} (2\alpha\beta)^{2z-1} \alpha d\alpha d\beta,$$

where we have introduced new variables $\alpha = \sqrt{s}, \beta = \sqrt{t}$. To this expression, we add the similar expression by permuting α, β. This

1.3 The Logarithmic Derivative of the Gamma Function

gives the more symmetric expression

$$2^{2z-1}\Gamma(z)\Gamma(z+\frac{1}{2}) = 2\int_0^\infty\int_0^\infty e^{-(\alpha^2+\beta^2)}(2\alpha\beta)^{2z-1}(\alpha+\beta)d\alpha d\beta$$

$$= 4\int_0^\infty\int_0^\infty e^{-(\alpha^2+\beta^2)}(2\alpha\beta)^{2z-1}(\alpha+\beta)d\alpha d\beta,$$

where the last integral is over the sector $\sigma; 0 \leqslant \alpha < \infty, 0 \leqslant \beta \leqslant \alpha$. Introducing new variable

$$u = \alpha^2 + \beta^2, \qquad v = 2\alpha\beta,$$

we find that

$$2^{2z-1}\Gamma(z)\Gamma(z+\frac{1}{2}) = \int_0^\infty v^{2z-1}dv \int_0^\infty \frac{e^{-u}}{\sqrt{u-v}}du$$

$$= 2\int_0^\infty e^{-v}v^{2z-1}dv \int_0^\infty e^{-w^2}dw = \sqrt{\pi}\,\Gamma(2z).$$

This result can be extended to arbitrary complex values $z \neq 0, -\frac{1}{2}, -1, -\frac{3}{2}, \ldots$ by using the principle of analytic continuation. We now use (1.6), and noting that $\Gamma(z) = 1$, we find by mathematical induction that

$$\Gamma(n+1) = n!, \qquad n = 0, 1, 2\ldots \tag{1.11}$$

Putting $z = \frac{1}{2}$, we obtain

$$\Gamma(\frac{1}{2}) = \int_0^\infty e^{-t}t^{-1/2}dt = 2\int_0^\infty e^{-u^2}du = \sqrt{\pi}, \tag{1.12}$$

and then (1.6) implies

$$\Gamma(n+\frac{1}{2}) = \frac{1.3.5\ldots\ldots(2n-1)}{2^n}\sqrt{\pi}, \qquad n = 1, 2, \ldots \tag{1.13}$$

1.3 The Logarithmic Derivative of the Gamma Function

The theory of the gamma function is intimately related to the theory of another special function, the logarithmic derivative of $\Gamma(z)$

$$\psi(z) = \frac{\Gamma'(z)}{\Gamma(z)}. \tag{1.14}$$

Since $\Gamma(z)$ is a meromorphic function with no zeros, $\psi(z)$ can have no singular points other than the poles $z = -n$ ($n = 0, 1, 2....$) of $\Gamma(z)$. It follows from (1.5) that $\psi(z)$ has the representation

$$\psi(z) = -\frac{1}{z+n} + \Omega(z+n) \tag{1.15}$$

in a neighbourhood of the point $z = -n$ and $\psi(z)$ is a meromorphic function with simple poles at the point $z = 0, -1, -2...$

The function $\psi(z)$ satisfied relations obtained from expression (1.6)–(1.8) by taking logarithmic derivatives. In this way, we find that

$$\psi(z+1) = \frac{1}{z} + \psi(z), \tag{1.16}$$

$$\psi(1-z) - \psi(z) = \pi \cot \pi z, \tag{1.17}$$

$$\psi(z) + \psi(z + \frac{1}{2}) + 2\log 2 = 2\psi(2z). \tag{1.18}$$

These formulas can be used to calculate $\psi(z)$ for special values of z. For example, writing

$$\psi(1) = \Gamma'(1) = -\Upsilon, \tag{1.19}$$

where $\Gamma - 0.57721566...$ is Euler's constant, and we obtain

$$\psi(n+1) = -\gamma + \sum_{k=1}^{n} \frac{1}{k}, \qquad n = 1, 2........ \tag{1.20}$$

1.3 The Logarithmic Derivative of the Gamma Function

Moreover, substituting $z = \frac{1}{2}$ into (1.18), we find that

$$\psi(\frac{1}{2}) = -\gamma - 2log2, \tag{1.21}$$

which then gives

$$\psi(n + \frac{1}{2}) = -\gamma - 2log2 + 2\sum_{k=1}^{n}\frac{1}{2k-1}, \quad n = 1, 2, \ldots \tag{1.22}$$

The function $\psi(z)$ has simple representations in the form of definite integrals involving the variable z as a parameter. To derive these representations, we first note that

$$\Gamma'(z) = \int_0^\infty e^{-t}t^{z-1}logt\,dt, \quad \operatorname{Re} z > 0. \tag{1.23}$$

If we replace the logarithm in the integrand by

$$logt = \int_0^\infty \frac{e^{-x} - e^{-xt}}{x}dx, \quad \operatorname{Re} t > 0, \tag{1.24}$$

we find that

$$\Gamma'(z) = \int_0^\infty \frac{dx}{x}\int_0^\infty (e^{-x} - e^{-xt})e^{-t}t^{z-1}dt,$$

$$= \int_0^\infty \frac{dx}{x}[e^{-x}\Gamma(z) - \int_0^\infty e^{-t(x+1)}t^{z-1}dt].$$

Introducing the new variable of integration $u = t(x + 1)$, we find that the integral in brackets equals $(x + 1)^{-z}\Gamma(z)$. This leads to the following integral representation of $\psi(z)$:

$$\psi(z) = \int_0^\infty [e^{-x} - \frac{1}{(x+1)^z}]\frac{dx}{x}, \quad \operatorname{Re} z > 0. \tag{1.25}$$

8 The Gamma Function

To obtain another integral representation of $\psi(z)$, we write (1.25) in the form

$$\psi(z) = \lim_{\delta \to 0} \left[\int_0^\infty e^{-x} - \frac{1}{(x+1)^2} \right] \frac{dx}{x}$$

$$= \lim_{\delta \to 0} \left[\int_0^\infty \frac{e^{-x}}{x} dx - \int_0^\infty \frac{1}{(x+1)^2 x} \right],$$

and change the variable of integration in the second integral by setting $x + 1 = e^t$. This gives

$$\psi(z) = \lim_{\delta \to 0} \left[\int_0^\infty \frac{e^t}{t} dt - \int_{\log(1+\delta)}^\infty \frac{e^{-tz}}{1 - e^{-t}} dt \right]$$

$$\lim_{\delta \to 0} \left[\int_{\log(1+\delta)}^\infty \left(\frac{e^{-t}}{t} - \frac{e^{-tz}}{1-e^{-t}} \right) dt - \int_{\log(1+\delta)}^\infty \frac{e^{-tz}}{1 - e^{-t}} dt \right],$$

and therefore, since the second integral integral approaches zero as $\delta \to 0$,

$$\psi(z) = \int_0^\infty \left(\frac{e^t}{t} - \frac{e^{-tz}}{1 - e^{-t}} \right) dt, \qquad \text{Re } z > 0. \qquad (1.26)$$

Setting $z = 1$ and subtracting the result from (1.26), we find that

$$\psi(z) = -\gamma + \int_0^\infty \frac{e^{-t} - e^{tz}}{1 - e^{-t}} dt, \qquad \text{Re } z > 0, \qquad (1.27)$$

or

$$\psi(z) = -\gamma + \int_0^1 \frac{1 - x^{z-1}}{1 - x^-} dx, \qquad \text{Re } z > 0, \qquad (1.28)$$

1.4 Asymptotic Representation of the Gamma Function for Large |z|

where we have introduced the new variable of integration $x = e^{-t}$. From expression (1.28), we can deduce an important representation of $\psi(z)$ as an analytic expression valid for all $z \neq 0, -1, -2, \ldots$. To obtain this representation, we substitute the power series expansion

$$(1-x)^{-1} = 1 + x + x^2 + x^3 + \ldots x^n \ldots, \quad 0 \leq x < 1$$

into (1.28), and the result is

$$\psi(z) = -\gamma + \sum_{n=0}^{\infty} \left(\frac{1}{n+1} - \frac{1}{n+z} \right). \tag{1.29}$$

The series (1.29), whose terms are analytic functions for $z \neq 0, -1, -2, \ldots$ is uniformly convergent in the region defined by the inequalities

$$|z + n| \geq \delta > 0, \quad n = 0, 1, 2 \ldots \text{and } |z| < a,$$

since

$$\left| \frac{1}{n+1} - \frac{1}{n+z} \right| < \frac{a+1}{(n+1)(n-a)}$$

for $n \geqslant N > a$, and the series

$$\sum_{n=N}^{\infty} \frac{a+1}{(n+1)(n-a)}$$

converges. Therefore, since δ is arbitrarily small and a arbitrarily small and a arbitrarily large. The following infinite product representation of the gamma function:

$$\frac{1}{\Gamma(z+1)} = e^{\gamma z} \sum_{n=1}^{\infty} e^{-z/n} \left(1 + \frac{z}{n}\right). \tag{1.30}$$

This formula can be made the starting point for the theory of the gamma function.

1.4 Asymptotic Representation of the Gamma Function for Large |z|

To describe the behavior of a given function $f(z)$ as $|z| \to \infty$ within a sector $\alpha \leqslant arg\, z \leqslant \beta$, it is, in many cases, sufficient to derive an expression of the form

$$f(z) = \phi(z)[1 + r(z)]. \tag{1.31}$$

Here, $\phi(z)$ is a function of a simpler structure than $f(z)$, and $r(z)$ converges uniformly to zero as $|z| \to \infty$ within the given sector. Formulas of this type are called asymptotic representations of $f(z)$ for large $|z|$. It follows from (1.31) that the ratio $f(z)\phi(z)$ converges to unity as $|z| \to \infty$; the two functions $f(z)$ and $\phi(z)$ are asymptotically a fact we indicates by writing

$$f(z) \approx \phi(z), \qquad |z| \to \infty, \qquad \alpha \leqslant arg\, z \leqslant \beta. \tag{1.32}$$

An estimate of $|r(z)|$ gives the size of the error committed when $f(z)$ is replaced by $\phi(z)$ for large but finite $|z|$. We now look for a description of the behavior of the function $f(z)$ as $|z| \to \infty$, which is more exact than by (1.31). Suppose we derive the expression

$$f(z) = \phi(z)\left[\sum_{n=0}^{N} a_n z^{-n} + r_N(z)\right], \qquad a_o = 1, \quad N = 1, 2, \ldots, \tag{1.33}$$

where $z^N r_N(z)$ converges uniformly to zero as $|z| \to \infty$, $\alpha \leqslant arg\, z \leqslant \beta$. Then we write

$$f(z) \approx \sum_{n=0}^{N} a_n z^{-n}, \qquad |z| \to \infty, \qquad \alpha \leqslant arg\, z \leqslant \beta, \tag{1.34}$$

and the right-hand side is called an asymptotic series or asymptotic expansion of $f(z)$ for large $|z|$. It should be noted that this definition does not exist that the given series converges in the ordinary sense, and, on the contrary, the series will usually diverge. Asymptotic series are very useful, since, by taking a finite number of terms, we can obtain an arbitarily good approximation to the function $f(z)$ for sufficiently large $|z|$.

1.5 Definite Integrals Related to the Gamma Function

The class of integrals that can be expressed in terms of the gamma function is very large. Here, we consider few examples with the intent of deriving some formulas that will be needed later.

Our first result is the expression

$$\int_0^\infty e^{-pt} t^{z-1} dt = \frac{\Gamma(z)}{p^z}, \qquad \text{Re } p > 0, \qquad \text{Re } z > 0, \tag{1.35}$$

which is easily proved for positive real p by making the change of variables $s = pt$, and then using the integral

$$\Gamma(z) = \int_0^\infty e^{-t} t^{z-1} dt, \qquad \text{Re } z \succ 0.$$

The extension of (1.35) to arbitrary complex p with Re $p > 0$ is accomplished by using the principle of analytic continuation. Now consider the integral

$$B(x, y) = \int_0^1 t^{x-1} (1-t)^{y-1} dt, \qquad \text{Re } x > 0, \qquad \text{Re } y > 0, \tag{1.36}$$

known as the beta function. It is easy to see that (1.36) represents an analytic function in each of the complex variables x and y. If we introduce the new variable on integration $u = t/(1-t)$, then (1.36) becomes

$$B(x,y) = \int_0^\infty \frac{u^{x-1}}{(1+u)^{x+y}} du, \qquad \text{Re } x > 0, \quad \text{Re } y > 0. \tag{1.37}$$

Setting $p = 1+u$, $z = x+y$ in (1.35), we find that

$$\frac{1}{(1+u)^{x+y}} = \frac{1}{\Gamma(x+y)} \int_0^\infty e^{-(1+u)t} t^{x+y-1} dt, \tag{1.38}$$

and substituting the result into (1.37), we obtain

$$B(x,y) = \frac{1}{\Gamma(x+y)} \int_0^\infty e^{-t} t^{x+y-1} dt \int_0^\infty e^{-ut} u^{x-1} du$$

$$= \frac{\Gamma(x)}{\Gamma(x+y)} \int_0^\infty e^{-t} t^{y-1} dt = \frac{\Gamma(x)\Gamma(y)}{\Gamma(x+y)} \tag{1.39}$$

Thus, we have derived the formula

$$B(x,y) = \frac{\Gamma(x)\Gamma(y)}{\Gamma(x+y)}, \tag{1.40}$$

relating the beta function to the gamma function, which can be used to derive all the properties of the beta function.

1.6 Exercises

1. Prove that

$$\int_0^{\pi/2} \cos^v\theta\, d\theta = \int_0^{\pi/2} \sin^v\theta\, d\theta = \frac{\sqrt{\pi}\,\Gamma(\frac{v+1}{2})}{2\Gamma(\frac{v}{2}+1)}, \qquad \text{Re } v - 1,$$

$$\int_0^{\pi/2} \cos^v\theta \sin^v\theta\, d\theta = \frac{1\Gamma(\frac{\mu+1}{2})\Gamma(\frac{v+1}{2})}{2\Gamma(\frac{\mu+v}{2}+1)}, \quad \text{Re}\,\mu > -1, \quad \text{Re}\,v > -1.$$

2. Verify the formula

$$\Gamma(3z) = \frac{3^{3z-1/2}}{2\pi}\Gamma(z)\Gamma(z+\frac{1}{3})\Gamma(z+\frac{2}{3})$$

3. Prove that

$$|\Gamma(iy)|^2 = \frac{\pi}{y\sinh \pi y}, \qquad |\Gamma(\frac{1}{2}+iy)|^2 = \frac{\pi}{y\cosh \pi y},$$

for real y.

4. Derive the formula

$$3\Psi(3z) = \Psi(z) + \Psi(z+\frac{1}{3}) + \Psi(z+\frac{2}{3}) + 3\log 3.$$

5. Derive the asymptotic formula

$$|\Gamma(x+iy)| = \sqrt{2\pi}\, e^{-1/2\pi|y|}|y|^{x-1/2}[1+r(x,y)],$$

where as $|t| \to \infty$, $r(x,y) \to 0$ uniformly in the strip $|x| \leqslant \alpha$ (α is a constant).

6. Derive the expressions

$$\gamma(z+1,\alpha) = z\gamma(z,\alpha) - e^{-\alpha}\alpha^z$$

$$\Gamma(z+1,\alpha) = z\Gamma(z,\alpha) - e^{-\alpha}\alpha^z$$

2
The Probability Integral and Related Functions

2.1 The Probability Integral and its Basic Properties

The probability integral is meant to be the function defined for any complex z by the integral

$$\phi(z) = \frac{2}{\sqrt{\pi}} \int_0^z e^{-t^2} dt, \qquad (2.1)$$

calculated along an arbitrary path joining the origin to the point $t = z$. The form of this path does not matter. The integrand is an entire function of the complex variable t. We can assume that the integration is along the line segment joining the point $t = 0$ and $t = z$. According to a familiar theorem of complex variable theory, $\phi(z)$ is an entire function and hence can be expanded in convergent power series for any value of z. To find this expansion, we need to only replace e^{-t^2} by its power series in (2.1) and then integrate the term obtaining

$$\phi(z) = \frac{2}{\sqrt{\pi}} \int_0^z \sum_{k=0}^{\infty} \frac{(-1)^k t^{2k}}{k!} dt = \frac{2}{\sqrt{\pi}} \sum_{k=0}^{\infty} \frac{(-1)^k z^{2k+1}}{k!(2k+1)}, \qquad |z| < \infty. \qquad (2.2)$$

16   The Probability Integral and Related Functions

It follows from (2.2) that $\phi(z)$ is an odd function of z. For real values of z, $\phi(z)$ is a real monotonically increasing function. At zero, we have $\phi(0) = 0$, and as z increases, $\phi(z)$ rapidly approaches the limiting value $\phi(\infty) = 1$, since

$$\int_0^\infty e^{-t^2} dt = \frac{\sqrt{\pi}}{2}. \qquad (2.3)$$

The difference between $\phi(z)$ and this limit can be written in the form

$$1 - \phi(z) = \frac{2}{\sqrt{\pi}} \int_0^\infty e^{-t^2} dt - \frac{2}{\sqrt{\pi}} \int_0^z e^{-t^2} dt = \frac{2}{\sqrt{\pi}} \int_0^\infty e^{-t^2} dt. \qquad (2.4)$$

The probability integral is encountered in many branches of applied mathematics, probability theory of errors, the theory of heat conduction, and various branches of mathematical physics. The two functions related to the probability integral, the error function

$$Erf\ z = \int_0^z e^{-t^2} dt = \frac{\sqrt{\pi}}{2} \phi(z), \qquad (2.5)$$

and its complement

$$Erfc\ z = \int_z^\infty e^{-t^2} dt = \frac{\sqrt{\pi}}{2} [1 - \phi(z)]. \qquad (2.6)$$

Many more complicated integrals can be expressed in terms of the probability integral. For example, by differentiation of the parameter z, it can be shown that

$$\frac{2}{\pi} \int_0^\infty \frac{e^{-zt^2}}{1+t^2} dt = e^z [1 - \phi(\sqrt{z})]. \qquad (2.7)$$

2.2 Asymptotic Representation of Probability Integral for Large |z|

To find an asymptotic representation of the function $\phi(z)$ for large $|z|$, We apply repeated integration by parts to the given integral

$$\int_z^\infty e^{-t^2}dt = -\frac{1}{2}\int_z^\infty \frac{1}{t}d(e^{-t^2}) = \frac{e^{-z^2}}{2z} - \frac{1}{2}\int_z^\infty \frac{e^{-t^2}}{t^2}dt$$

$$= \frac{e^{-z^2}}{2z} - \frac{e^{-z^2}}{2^2 z^3} + \frac{1.3}{2^2}\int_z^\infty \frac{e^{-t^2}}{t^4}dt$$

$$= e^{-z^2}\left[\frac{1}{2z} - \frac{1}{2^2 z^3} + \frac{1.3}{2^3 z^5} - \frac{1.3.5}{2^4 z^7} + \ldots\ldots\right.$$

$$\left. +(-1)^n\frac{1.3\ldots\ldots(2n-1)}{2^{n+1} z^{2n+1}}\right]$$

$$+ (-1)^{n+1}\frac{1.3\ldots\ldots(2n-1)}{2^{n+1} z^{2n+1}}\int_z^\infty \frac{e^{-t^2}}{t^{2n+1}}dt.$$

It follows that

$$1 - \phi(z) = \frac{e^{-z^2}}{\sqrt{\pi}z}\left[1 + \sum_{k=1}^{n}(-1)^k\frac{1.3\ldots(2k-1)}{(2z^2)^k} + r_n(z)\right], \quad (2.8)$$

where

$$r_n(z) = (-1)^{n+1}\frac{1.3\ldots\ldots(2n+1)}{2^n}ze^{z^2}\int_z^\infty \frac{e^{-t^2}}{t^{2n+2}}dt. \quad (2.9)$$

where δ is an arbitrarily small positive number, and choose the path of integration in (2.9) to be the infinite line segment beginning at the point $t = z$ and parallel to the real axis. If $z = x + iy = re^{i\Phi}$, then this segment has the equation $t = u + iy$ $(x \leqslant u < \infty)$, and on the segment, we have

$$|e^{z^2-t^2}| = e^{x^2-u^2}, \qquad |t|^{-(2n+3)} \leqslant |z|^{-(2n+3)}, \qquad |t| \leqslant u\sec\phi.$$

so that

$$|r_n(z)| < \frac{1.3......(2n+1)}{2^n|z|^{2n+2}}\sec\phi \int_z^\infty e^{x^2-u^2} u\, du = \frac{1.3......(2n+1)}{(2|z|^2)^{n+1}}\sec\phi,$$

which implies

$$|r_n(z)| \leqslant \frac{1.3......(2n+1)}{(2|z|^2)^{n+1}}\sec\phi \leqslant \frac{1.3......(2n+1)}{(2|z|^2)^{n+1}\sin\delta}. \qquad (2.10)$$

It follows from (2.10) that as $|z| \to \infty$ the product $z^{2n}r_n(z)$ converges uniformly to zero in the indicated sector,

$$1 - \phi(z) \approx \frac{e^{-z^2}}{\sqrt{\pi}z}\left[1 + \sum_{k=1}^n (-1)^n \frac{1.3.....(2n-1)}{(2z^2)^n}\right], \qquad (2.11)$$

$$|z| \to \infty, \qquad |arg\, z| \leqslant \frac{\pi}{2} - \delta.$$

Thus, the series on the right is the asymptotic series of the function $1-\phi(z)$, and a bound on the error committed in approximating $1-\phi(z)$ by the sum of a finite number of terms of the series is given by (2.10). For positive real z, this error does not exceed the first neglected term is absolute value.

2.3 The Probability Integral of Imaginary Argument

In the applications, one often encounters the case where the argument of the probability integral is a complex. We now examine the particular simple case where $z = ix$ is a purely imaginary. Choosing a segment

2.3 The Probability Integral of Imaginary Argument

of the imaginary axis as the path of integration, and making the substitution $t = iu$, we find from (2.11) that

$$\frac{\Phi(ix)}{i} = \frac{2}{\sqrt{\pi}} \int_0^x e^{u^2} du. \tag{2.12}$$

The integration in the right increases without limit as $x \to \infty$, and, therefore, it is more convenient to consider the function

$$F(z) = e^{-z^2} \int_0^z e^{u^2} du, \tag{2.13}$$

which remains bounded for all real z. In the general case of complex z, $F(z)$ is an entire function, and the choice of the path of integration in (2.13) is complete. To expand $F(z)$ in power series, we note that $F(z)$ satisfies the linear differential equation

$$F'(z) + 2zF(z) = 1, \tag{2.14}$$

with initial condition $F(0) = 0$. Substituting the series

$$F(z) = \sum_{k=0}^{\infty} a^k z^k$$

into (2.14), and comparing coefficients if identical powers of z, we obtain recurrence relation

$$a_0 = 0, \quad a_1 = 1, \quad (k+1=1)a_{k+1} + 2a_{k-1} = 0.$$

After some simple calculations, this leads to the expansion

$$F(z) = \sum_{k=0}^{\infty} \frac{(-1)^k z^{2k+1}}{1.3.....(2k+1)}, \quad |z| < \infty. \tag{2.15}$$

To study the behavior of $F(z)$ as $z \to \infty$ for real z, we apply L'Hospital's rule twice to the ratio

$$\frac{2z \int_0^z e^{u^2} du}{e^{z^2}},$$

and then use (2.14) to deduce that

$$\lim_{z\to\infty} 2zF(z) = 1,$$

and

$$F(z) \approx \frac{1}{2z}, \qquad z \to \infty. \qquad (2.16)$$

The maximum of the function occurs at $z = 0.924...$ and equals $F_{max} = 0.541...$

The function $F(z)$ comes up in the theory of propagation of electromagnetic waves along the earth's surface, and in other problems of mathematical physics.

2.4 The Probability Fresnel Integrals

Another interesting case from the stand point of the application occurs, when the argument of the probability integral is the complex number

$$z = \sqrt{ix} = \frac{x}{\sqrt{2}}(1+i),$$

where x is real. In this case, we choose the path of the integration in (2.11) to be a segment of the bisector of the angle between the real and imaginary axes. Then, using the formula $t = \sqrt{iu}$ to introduce the new variable u, we find from (1.1) that

$$\frac{\phi(\sqrt{ix})}{(\sqrt{i})} = \frac{2}{\sqrt{\pi}} \int_0^x e^{-iu^2} du = \frac{2}{\sqrt{\pi}} \int_0^x \cos u^2 du - i\frac{2}{\sqrt{\pi}} \int_0^x \sin u^2 du. \qquad (2.17)$$

The integrals on the right can be expressed in terms of the functions

$$C(z) = \int_0^x \cos\frac{\pi t^2}{2} dt, \qquad S(z) = \int_0^x \sin\frac{\pi t^2}{2} dt, \qquad (2.18)$$

2.4 The Probability Fresnel Integrals

where the integration is along any path joining the origin to the point $t = z$. The function $C(z)$ and $S(z)$ are known as the Fresnel integrals. Since the integrands in (2.28) are functions of the complex variable t, the choice of the path of integration does not matter, and both $C(z)$ and $S(z)$ are entire functions of z. For real $z = x$, the Fresnel integrals are real. Both $C(x)$ and $S(x)$ vanish for $x = 0$, and have an oscillatory character, as follows from the expressions

$$C'(x) = \cos\frac{\pi x^2}{2}, \qquad S'(x) = \sin\frac{\pi x^2}{2},$$

which shows that $C(x)$ has extrema at $x = \pm\sqrt{2n+1}$, which $S(x)$ has extrema at $x = \pm\sqrt{2n}$ ($n = 0, 1, 2...$). The largest maxima are $C(1) = 0.779893...$ and $S(\sqrt{2}) = 0.713972..$, respectively. As $x \to \infty$, each of the functions approaches the limit

$$C(\infty) = S(\infty) = \frac{1}{2},$$

as implied by the familiar expression

$$\int_0^\infty \cos t^2 dt = \int_0^\infty \sin t^2 dt = \frac{\sqrt{\pi}}{2\sqrt{2}}. \qquad (2.19)$$

Replacing the trigonometric functions in the integrands in (2.58) by their power series expansions, and integrating term by term, we obtain the following series expansions for the Fresnel integrals

$$C(z) = \int_0^z \sum_{k=0}^\infty \frac{(-1)^k}{(2k)!} \left(\frac{\pi}{2}\right)^{2k} \frac{z^{4k+1}}{4k+1},$$

$$S(z) = \int_0^z \sum_{k=0}^\infty \frac{(-1)^k}{(2k+1)!} \left(\frac{\pi t^2}{2}\right)^{2k+1} dt = C(z)$$

$$= \sum_{k=0}^\infty \frac{(-1)^k}{(2k+1)!} \left(\frac{\pi}{2}\right)^{2k+1} \frac{z^{4k+3}}{4k+3}.$$

The relation between the Fresnel integrals and the probability integral is given by the formula

$$C(z) \pm iS(z) = \int_0^z e^{\pm \pi t^2/2} dt = \sqrt{\frac{2}{\pi}} e^{\pm \pi i/4} \int_0^{\sqrt{\pi/2} z e^{\pm \pi i/4}} e^{-u^2} du$$

$$= \frac{1}{\sqrt{2}} e^{\pm \pi i/4} \phi\left(\sqrt{\frac{\pi}{2}} z e^{\pm \pi i/4}\right), \qquad (2.20)$$

which implies

$$S(z) = \frac{1}{2i\sqrt{2}} \left[e^{\pi i/4} \Phi\left(\sqrt{\frac{\pi}{2}} z e^{-\pi i/4}\right) - e^{-\pi i/4} \Phi\left(\sqrt{\frac{\pi}{2}} z e^{\pi i/4}\right) \right], \qquad (2.21)$$

$$C(z) = \frac{1}{2\sqrt{2}} \left[e^{\pi i/4} \Phi\left(\sqrt{\frac{\pi}{2}} z e^{-\pi i/4}\right) + e^{-\pi i/4} \Phi\left(\sqrt{\frac{\pi}{2}} z e^{\pi i/4}\right) \right]. \qquad (2.22)$$

Using (2.21) and (2.22), we can derive the properties of $C(z)$ and $S(z)$ from the corresponding properties of $C(z)$ and $S(z)$ of the probability integral. The following asymptotic representations of the Fresnel integrals, valid for large $|z|$ in the sector $|arg z| \leq \frac{\pi}{4} - \delta$,

$$C(z) = \frac{1}{2} - \frac{1}{\pi z} \left[B(z) \cos \frac{\pi z^2}{2} - A(z) \sin \frac{\pi z^2}{2} \right], \qquad (2.23)$$

$$S(z) = \frac{1}{2} - \frac{1}{\pi z} \left[A(z) \cos \frac{\pi z^2}{2} + B(z) \sin \frac{\pi z^2}{2} \right], \qquad (2.24)$$

where

$$A(z) = \sum_{k=0}^{k=n} \frac{(-1)^k \alpha_{2k}}{(\pi z^2)^{2k}} + O(|z|^{-4N-4}),$$

$$B(z) = \sum_{k=0}^{k=n} \frac{(-1)^k \alpha_{2k+1}}{(\pi z^2)^{2k+1}} + O(|z|^{-4N-6}),$$

$$\alpha_k = 1.3 \ldots (2k-1), \qquad \alpha_0 = 1.$$

The Fresnel integrals come up in various branches of physics and engineering, diffraction, and theory of vibrations. Many integrals of more complicated type can be expressed in terms of the functions $C(z)$ and $S(z)$.

2.5 Application to Probability Theory

Normal or Gaussian random variable with mean m and standard deviation σ is meant to be a random variable ξ such that the probability of ξ lying in the interval $[x, x + \mathrm{d}x]$ is given by the expression

$$\frac{1}{\sqrt{2\pi}\sigma} e^{-(x-m)^2/2\sigma^2} \mathrm{d}x. \qquad (2.25)$$

The probability

$$P\{a \leqslant \xi - m \leqslant b\} \qquad (2.26)$$

that $\xi - m$ lies in the interval $[a, b]$ is just the integral

$$\frac{1}{\sqrt{2\pi}\sigma} \int_{a+m}^{b+m} e^{-(x-m)^2/2\sigma^2} \mathrm{d}x = \frac{1}{\sqrt{\pi}} \int_{a/\sqrt{2}\sigma}^{a/\sqrt{2}\sigma} e^{-t^2} \mathrm{d}t$$

$$= \frac{1}{2}\left[\Phi\left(\frac{b}{\sqrt{2}\sigma}\right) - \Phi\left(\frac{a}{\sqrt{2}\sigma}\right)\right], \qquad (2.27)$$

where $\Phi(x)$ is the probability integral. Setting $a = -\delta$, $b = \delta$, we obtain the probability that $|\xi - m|$ does not exceeds δ :

$$P\{|\xi - m| \leqslant \delta\} = \Phi(\frac{\delta}{\sqrt{2}\sigma}). \qquad (2.28)$$

The the probability that $|\xi - m|$ exceed δ is just

$$P\{|\xi - m| > \delta\} = 1 - \Phi(\frac{\delta}{\sqrt{2}\sigma}). \qquad (2.29)$$

The value $\delta = \delta_p$ for which (2.28) and (2.29) are equal is called the probable error, and it clearly satisfies the equation

$$\Phi(\frac{\delta_p}{\sqrt{2}\sigma}) = \frac{1}{2}.$$

Using a table of the function $\Phi(x)$ to solve the equation, we find that

$$\delta_p = 0.67449\sigma. \tag{2.30}$$

2.6 Application to the Theory of Heat Conduction

Consider the following problem in the theory of heat conduction. An object occupying the half-surface $x \geq 0$ is initially heated to temperature T_o. It then cools off by radiating heat through its surface $x = 0$ into the surrounding medium which is at zero temperature. We want to find the temperature $T(x,t)$ of the object as a function of position x and time t. Let the object have thermal conductivity k, heat capacity c, density ρ, and emissivity λ, and let $\tau = kt/c\rho$. Then our problem reduces to the solution of the equation of heat conduction

$$\frac{\partial T}{\partial \tau} = \frac{\partial^2 T}{\partial x^2}, \tag{2.31}$$

subject to the initial condition

$$T\,|_{\tau=0} = T_0 \tag{2.32}$$

and the boundary conditions

$$(\frac{\partial T}{\partial \tau} - hT)|_{x=0} = 0, \qquad T|_{x \to \infty}, \tag{2.33}$$

where $h = \lambda/k > 0$.

2.6 Application to the Theory of Heat Conduction

To solve the problem, we introduce the Laplace transformation $\bar{T} = \bar{T}(x, p)$ of $T = T(x, \tau)$, defined by the expression

$$\bar{T} = \int_0^\infty e^{-p\tau} T d\tau, \qquad \operatorname{Re} p > 0. \tag{2.34}$$

A system of equations determining $\bar{T}$ can be obtained from (2.31) to (2.33) if we multiply the first and third equations by $e^{-p\tau}$ and integrate from 0 to ∞, taking the second equation into account. The result is

$$\frac{d^2 \bar{T}}{dx^2} = p\bar{T} - T_o,$$

$$\frac{d\bar{T}}{dx} - h\bar{T}\Big|_{x=0}, \qquad \bar{T}\Big|_{x\to\infty} = \frac{T_o}{p}. \tag{2.35}$$

Equation (2.35) has the solution

$$\bar{T} = \frac{T_o}{p}\left(1 - \frac{h}{h + \sqrt{p}} e^{-\sqrt{p}x}\right), \qquad \operatorname{Re} p > 0, \quad \operatorname{Re}\sqrt{p} > 0. \tag{2.36}$$

We can now solve for T by inverting (2.34). This can be done by Laplace transform or Fourier–Millin inversion theorem, which state that

$$T = \frac{1}{2\pi i} \int_{a-i\infty}^{a+i\infty} e^{p\tau} \bar{T} dp, \tag{2.37}$$

where a is constantly greater than the real part of all the singular points of $\bar{T}$.

The quantity of greatest interest is the surface temperature of the object. If we, put $x = 0$ in (2.36), we find that

$$\bar{T}\Big|_{x=0} = \frac{T_o}{\sqrt{p}(\sqrt{p} + h)} = T_o\left(\frac{1}{p - h^2} - \frac{h}{p - h^2}\frac{1}{\sqrt{p}}\right). \tag{2.38}$$

The simplest way to solve (2.38) for the original function $T|_{x=0}$ is to use the convolution theorem, which states that if f_1 and f_2 are the Laplace transforms of f_1 and f_2, then $\dot{f} = \dot{f}_1 \dot{f}_2$ is the Laplace transform of the function

$$f(\tau) = \int_0^\tau f_1(t) f_2(\tau - t) \mathrm{d}t. \tag{2.39}$$

Since it easily verified that

$$\dot{f}_1 = \frac{h}{\sqrt{p}}, \qquad \dot{f}_2 = \frac{1}{p - h^2}$$

are the Laplace transform of

$$f_1 = \frac{h}{\sqrt{\pi \tau}}, \qquad f_2 = e^{h^2 \tau},$$

(2.39) implies

$$T|_{x=0} = T_o \left(e^{h^2 \tau} - \frac{h}{\sqrt{\pi}} \int_0^\infty \frac{\mathrm{d}t}{\sqrt{t}} \right) = T_o e^{h^2 \tau} \left(1 - \frac{2}{\sqrt{\pi}} \int_0^{h\sqrt{\tau}} e^{-s^2} \mathrm{d}s \right),$$

$$T|_{x=0} = T_o e^{h^2 \tau} \left[1 - \Phi(h\sqrt{\tau}) \right], \tag{2.40}$$

where $\Phi(x)$ is the probability integral. It follows from the asymptotic formula (2.11) that for large τ, the surface temperature falls off like $1/\sqrt{\tau}$:

$$T|_{x=0} \approx \frac{T_o}{h\sqrt{\pi \tau}}, \qquad \tau \to \infty. \tag{2.41}$$

The temperature inside the object ($x \neq 0$) can also be expressed in closed form in terms of the probability integral.

2.7 Application to the Theory of Vibrations

Consider an infinite rod of linear density ρ and Young's modulus E, lying along the positive x-axis. Let I be the moment of inertia of a cross section of the rod about a horizontal axis through the center of

2.7 Application to the Theory of Vibrations

mass of the section, and let $\tau = \sqrt{EI/\rho t}$. Suppose the end $x = 0$ satisfies a sliding condition while the end $x = \infty$ is clamped, and suppose a constant force Q is suddenly applied at the end $x = 0$. The displacement $u = u(x,t)$ at an arbitrary point $x \geqslant 0$ of the rod is described by the system of equation

$$\frac{\partial^2 u}{\partial \tau^2} + \frac{\partial^4 u}{\partial x^2} = 0,$$

$$u|_{\tau=0} = \frac{\partial u}{\partial \tau}\Big|_{\tau=0} = 0, \quad (2.42)$$

$$\frac{\partial u}{\partial x}\Big|_{x=0}, \quad \frac{\partial^3 u}{\partial x^3}\Big|_{x=0} = \frac{Q}{EI}, \quad u|_{x\to\infty} = 0, \quad \frac{\partial u}{\partial x}\Big|_{x\to\infty} = 0.$$

To solve this system, we use the Laplace transform, as in the preceding section. Writing

$$\bar{u} = \int_0^\infty e^{-p\tau} u \, d\tau, \qquad \operatorname{Re} p > 0, \quad (2.43)$$

we obtain the following equations for $\bar{u}$:

$$\frac{\partial^4 \bar{u}}{\partial x^4} + p^2 \bar{u} = 0,$$

$$\frac{\partial \bar{u}}{\partial x}\Big|_{x=0}, \quad \frac{\partial^3 \bar{u}}{\partial x^3}\Big|_{x=0} = \frac{Q}{EIp},$$

$$\bar{u}|_{x=0} = 0, \quad \frac{\partial \bar{u}}{\partial x}\Big|_{x\to\infty} = 0.$$

Simple calculation then shows that

$$\bar{u} = \frac{Q}{2EIp^2 i}\left(\frac{e^{-\sqrt{-pi}\,x}}{\sqrt{-pi}} - \frac{e^{-\sqrt{pi}\,x}}{\sqrt{pi}}\right), \quad \operatorname{Re} p > 0, \ \operatorname{Re} \sqrt{\pm pi} > 0. \quad (2.44)$$

To find u, we again use the convolution theorem,

$$\bar{f}_1 = \frac{Q}{EIp^2}, \qquad \bar{f}_2 = \frac{1}{2i}\left(\frac{e^{-\sqrt{-pi}\,x}}{\sqrt{-pi}} - \frac{e^{-\sqrt{pi}\,x}}{\sqrt{pi}}\right)$$

are the Laplace transforms of

$$f_1 = \frac{Q}{EI}\tau, \qquad f_2 = \frac{1}{2i}\left(\sin\frac{x^2}{4\tau} + \cos\frac{x^2}{4\tau}\right),$$

(2.39) implies

$$u = \frac{Q}{EI\sqrt{2\pi}} \int_0^\tau \left(\sin\frac{x^2}{4t} + \cos\frac{x^2}{4t}\right)\frac{\tau - t}{\sqrt{t}}dt = \frac{Q x \tau}{EI} f\left(\frac{x}{2\sqrt{\tau}}\right), \qquad (2.45)$$

where

$$f(x) = \frac{1}{\sqrt{2\pi}} \int_0^\infty (\sin y^2 + \cos y^2) \frac{1 - \frac{x^2}{y^2}}{y^2} dy. \qquad (2.46)$$

The function $f(x)$ can be expressed in terms of the Fresnel integrals $C(z)$ and $S(z)$, introduced in Section 2.4. Integrating (2.45) by parts twice, we find that

$$f(x) = \left(1 + \frac{2}{3}x^2\right)\left[\frac{1}{2} - C\left(\sqrt{\frac{2}{\pi}}x\right)\right] - \left(1 - \frac{2}{3}x^2\right)\left[\frac{1}{2} - S\left(\sqrt{\frac{2}{\pi}}x\right)\right]$$

$$+ \frac{2}{3\sqrt{2\pi}}\left[(1 + x^2)\frac{\sin x^2}{x} + (1 - x^2)\frac{\cos x^2}{x}\right]. \qquad (2.47)$$

2.8 Exercises

(1) Show that the function

$$\phi(z) = \frac{\sqrt{\pi}}{2} e^{z^2} \Phi(z)$$

satisfies the differential equation $\phi' - 2z\phi = 1$, and use this fact to derive the expansion

$$\Phi(z) = \frac{2z}{\sqrt{\pi}} e^{-z^2} \sum_{k=0}^\infty \frac{(2z^2)^k}{1 \cdot 3 \cdots (2k+1)}, \qquad |z| < \infty.$$

(2) Use integration by parts to show that

$$\int \Phi(x)\mathrm{d}x = x\Phi(x) + \frac{1}{\sqrt{\pi}}\mathrm{e}^{-x^2} + C.$$

(3) Let $\bar{\Phi}$ be the laplace transform of the probability integral,

$$\bar{\Phi}(p) = \int_0^\infty \mathrm{e}^{-px}\Phi(x)\mathrm{d}x.$$

Prove that

$$\bar{\Phi}(p) = \frac{1}{4}\mathrm{e}^{p^2/4}\left[1 - \Phi(\frac{p}{2})\right].$$

(4) Derive the integral representations

$$F(z) = \int_0^\infty \mathrm{e}^{-t^2}\sin 2zt\,\mathrm{d}t, \qquad \Phi(z) = \frac{2}{\pi}\int_0^\infty \mathrm{e}^{-t^2}\frac{\sin 2zt}{t}\mathrm{d}t.$$

(5) Derive the following integral representations for the square of the probability integral:

$$\Phi^2(z) = 1 - \frac{4}{\pi}\int_0^1 \frac{\mathrm{e}^{-z^2(1+t^2)}}{1+t^2}\mathrm{d}t,$$

$$[1 - \Phi(z)]^2 = \frac{4}{\pi}\int_1^\infty \frac{\mathrm{e}^{-z^2(1+t^2)}}{1+t^2}\mathrm{d}t, \qquad |arg z| \leqslant \frac{\pi}{4}.$$

(6) Derive the formulas

$$1 - \Phi(z) = \frac{2}{\sqrt{\pi}}\mathrm{e}^{-z^2}\int_0^1 \mathrm{e}^{-t^2-2zt}\mathrm{d}t,$$

$$[1 - \Phi(z)]^2 = \frac{4}{\sqrt{\pi}}\mathrm{e}^{-2z^2}\int_0^\infty \mathrm{e}^{-t^2-2\sqrt{2}zt}\Phi(t)\mathrm{d}t.$$

3

Spherical Harmonics Theory

3.1 Introduction

By spherical harmonics, we mean solutions of the linear differential equation

$$(1-z^2)u'' - 2zu' + \left[v(v+1) - \frac{\mu^2}{1-z^2}\right]u = 0, \quad (3.1)$$

where z is complex variable, and μ, v are parameters that can take arbitrary real or complex values. Equation(3.1) is encountered in mathematical physics when a system of orthogonal curvilinear coordinates to solve the boundary value problems of potential theory for certain special kinds of domains like the sphere, spheroid, and the simplest of these domains is the sphere, which gives rise to the term "spherical harmonics." In the spherical case, the variable z takes real values in the interval $(-1, 1)$, and the parameters μ and v are non-negative integers, but boundary value problems with more complicated geometries lead to the consideration of more general values of z, μ, and v. For most applications, it is sufficient to assume that z is either a real variable in the interval $(-1, 1)$ or a complex variable in the plane cut along the segment $[-\infty, 1]$, while v is an arbitrary real or complex number and $\mu = m$ is a non-negative integer ($m = 0, 1, 2, 3...$).

3.2 The Hypergeometric Equation and its Series Solution

Before considering the theory of spherical harmonics, it is appropriate to consider the problem of solving the linear differential equation

$$z(1-z)u'' + [\gamma - (\alpha+\beta+1)z]u' - \alpha\beta u = 0, \qquad (3.2)$$

where z is complex variable, and α, β, $and\,\gamma$ are parameters which can take various real or complex values. Equation (3.1) is called the hypergeometric equation, and contains as special cases many differential equations encountered in the applications. Reducing (3.1) to a standard form by dividing it by the coefficient of u'', we obtain an equation whose coefficients are analytic functions of z in the domain $0 < |z| < 1$ and have the point $z = 0$ as a simple pole or a regular point, depending on the values of the parameters α, β, $and\,\gamma$. It follows from the general theory of linear differential equations that (3.1) has a particular solution of the form

$$u = z^s \sum_{k=0}^{\infty} c_k z^k, \qquad (3.3)$$

where $c_o \neq 0, s$ is a suitable chosen number, and the power series converges for $|z| < 1$. Substituting (3.2) into (3.1), we find that

$$\sum_{k=0}^{\infty} c_k z^{s+k-1}(s+k)(s+k-1+\gamma) - \sum_{k=0}^{\infty} c_k z^{s+k}(s+k+\alpha)(s+k+\beta) = 0,$$
$$(3.4)$$

which gives the following system of equations for determining the exponent s and the coefficients c_k:

$$c_0 s(s-1+\gamma) = 0,$$

3.2 The Hypergeometric Equation and its Series Solution

$$c_k(s+k)(s+k-1+\gamma) - c_{k-1}(s+k-1+\alpha)(s+k-1+\beta) = 0,$$
$$k = 1, 2, ...$$

Solving the first equation, we obtain $s = 0$ or $s = 1 - \gamma$. Suppose $\gamma \neq 0, -1, -2, ...$ and choose $s = 0$. Then the coefficients c_k can be calculated from the recurrence relation

$$c_k = \frac{(k-1+\alpha)(k-1+\beta)}{k(k-1+\gamma)} c_{k-1}, \qquad k = 1, 2...,$$

if we set $c_o = 1$, this implies

$$c_k = \frac{(\alpha)_k (\beta)_k}{k! (\gamma)_k}, \qquad k = 0, 1, 2...,$$

where we have introduced the abbreviation

$$(\lambda)_o = 1, \quad (\lambda)_k = \lambda(\lambda+1)..........(\lambda+k-1), \quad k = 1, 2... \quad (3.5)$$

Thus, if $\gamma \neq 0, -1, -2, ...$, a particular solution of (3.2) is

$$u = u_1 = F(\alpha, \beta; \gamma; z) = \sum_{k=0}^{\infty} \frac{(\alpha)_k (\beta)_k}{k! (\gamma)_k} z^k, \qquad |z| < 1, \quad (3.6)$$

where the series on the right is known as the hypergeometric series. The convergence of this series for $|z| < 1$ follows from the general theory of linear differential equations. However, by using the ratio test, it can easily be proved without recourse to this theory that the radius of convergence of the series (3.6) is unity, except when the parameter α, β equals zero or a negative integer, in which case it reduces to polynomial.

Similarly, choosing $s = 1 - \gamma$ and assuming that $\gamma \neq 2, 3, 4, ...$, we obtain

$$c_k = \frac{(k-\gamma+\alpha)(k-\gamma+\beta)}{k(k+1-\gamma)} c_{k-1}, \qquad k = 1, 2, ..., \quad (3.7)$$

or

$$c_k = \frac{(1-\gamma+\alpha)_k(1-\gamma+\beta)_k}{k!(2-\gamma)_k}, \quad k = 0, 1, 2, ..., \quad (3.8)$$

If we set $c_o = 1$. Thus if $\gamma \neq 2, 3, 4, ...$, a particular solution of (3.2) is

$$u = u_2 = z^{1-\gamma} \sum_{k=0}^{\infty} \frac{(1-\gamma+\alpha)_k(1-\gamma+\beta)_k}{k!(2-\gamma)_k} z^k$$

$$= z^{1-\gamma} F(1-\gamma+\alpha, \; 1-\gamma+\beta; 2-\gamma; z), \quad |z| < 1, \; |argz| < \pi. \quad (3.9)$$

Therefore, if $\gamma \neq 0, 1, 2, ...$, the two solutions (3.8) and (3.9) exist simultaneously and are linearly independent. Then the general solution of (3.1) can be written in the following form:

$$u = AF(\alpha, \beta; \gamma; z) + Bz^{1-\gamma} F(1-\gamma+\alpha, 1-\gamma+\beta; 2-\gamma; z), \quad (3.10)$$

where $|z| < 1$, $|argz| < \pi$, and A, B are arbitrary constants. We can obtain a number of other differential equations whose solutions can be expressed in terms of hypergeometric series. Thus, for example, setting $z = t^2$, we arrive at the differential equation

$$t(1-t^2)\frac{d^2u}{dt^2} + 2[\gamma - \frac{1}{2} - (\alpha+\beta+\frac{1}{2})t^2]\frac{du}{dt} - 4\alpha\beta tu = 0, \quad (3.11)$$

with particular solutions

$$u = u_1 = F(\alpha, \beta; \gamma; t^2), \quad \gamma \neq 0, -1, -2, ..., \quad (3.12)$$

$$u = u_2 = t^{2-2\gamma} F(1-\gamma+\alpha, 1-\gamma+\beta; 2-\gamma; t^2)$$

$$|t| < 1, \quad |argt| < \pi, \quad \gamma \neq 2, 3, 4, ..., \quad (3.13)$$

which for nonintegral γ constitute a pair of linearly independent solutions of (3.11) in the domain $0 < |t| < 1$.

3.3 Legendre Functions

The simplest class of spherical harmonics consists of the Legendre polynomials. The solution of (3.2) for $\mu = 0$ and arbitrarily real or complex solutions of the equation

$$(1 - z^2)u'' - 2zu' + v(v+1)u = 0, \tag{3.14}$$

known as Legendre's equation. To determine these functions, we first note that (3.14) can be reduced to the hypergeometric equation by making suitable changes to variables. In particular, the substitution $t = \frac{1}{2}(1 - z)$ converts (3.14) into the equation

$$t(1 - t)u'' - 2zu' + v(v+1)u = 0, \tag{3.15}$$

which is a special case of (3.2) corresponding to

$$\alpha = -v, \quad \beta = v + 1, \quad \gamma = 1, \tag{3.16}$$

while the substitution $t = z^{-2}$, $u = z^{-v-1}v$ converts (3.14) into the equation

$$t(1-t)\frac{d^2v}{dt^2} + \left[\left(v + \frac{3}{2}\right) - \left(v + \frac{5}{2}\right)t\right]\frac{dv}{dt} - \left(\frac{v}{2} + 1\right)\left(\frac{v}{2} + \frac{1}{2}\right)v = 0, \tag{3.17}$$

which is a special case of (3.14) corresponding to

$$\alpha = \frac{v}{2} + 1, \quad \beta = \frac{v}{2} + \frac{1}{2}, \quad \gamma = v + \frac{3}{2}.$$

Therefore, it follows from the result of the section that two particular solutions of (3.14) are

$$u = u_1 = F(-v, v+1; 1; \frac{1-z}{2}), \quad |z - 1| < 2, \tag{3.18}$$

$$u = u_2 = \frac{\sqrt{\pi}\Gamma(v+1)}{\Gamma(v+\frac{3}{2})(2z)^{v+1}} F\left(\frac{v}{2}+1, \frac{v}{2}+\frac{1}{2}; v+\frac{3}{2}; \frac{1}{z^2}\right),$$

$$|z| > 1, \quad |arg\ z| < \pi, \quad v \neq -1, -2, \ldots\ldots, \tag{3.19}$$

where $F(\alpha, \beta; \gamma; z)$ is the hypergeometric series. These solutions are called the Legendre of degree v of the first and second kinds, denoted by $P_v(z)$ and $Q_v(z)$, respectively. Thus, we have

$$P_v(z) = F(-v, v+1; 1; \frac{1-z}{2}), \quad |z-1| < 2, \tag{3.20}$$

$$Q_v(z) = \frac{\sqrt{\pi}\Gamma(v+1)}{\Gamma(v+\frac{3}{2})(2z)^{v+1}} F\left(\frac{v}{2}+1, \frac{v}{2}+\frac{1}{2}; v+\frac{3}{2}; \frac{1}{z^2}\right),$$

$$|z| > 1, \quad |arg\ z| < \pi, \quad v \neq -1, -2, \ldots\ldots, \tag{3.21}$$

The functions $P_v(z)$ and $Q_v(z)$ are defined in certain restricted regions of the complex z-plane. To make the analytic continuation of $P_v(z)$, the legendre function of the first kind, we use the formula

$$\frac{2}{\pi}\int_0^{\pi/2} \sin^{2k}\phi d\phi = \frac{(\frac{1}{2})_k}{k!}, \quad k = 0, 1, 2 \tag{3.22}$$

to write (3.20) as

$$P_v(z) = \sum_{k=0}^{\infty} \frac{(-v)_k(v+1)_k}{(k!)^2} \left(\frac{1-z}{2}\right)^k$$

$$= \frac{2}{\pi}\sum_{k=0}^{\infty} \frac{(-v)_k(v+1)_k}{(\frac{1}{2})_k k!} \left(\frac{1-z}{2}\right)^k \int_0^{\pi/2} \sin^{2k}\phi d\phi. \tag{3.23}$$

To make the analytic continuation of $Q_v(z)$, the Legendre function of the second kind, we start the formula

$$\int_1^{\infty} \frac{dt}{t^{2k+v+\frac{3}{2}}\sqrt{t-1}} = \frac{\sqrt{\pi}\Gamma(v+1)(\frac{v}{2}+\frac{1}{2})_k(\frac{v}{2}+1)_k}{\Gamma(v+\frac{3}{2})(\frac{v}{2}+\frac{3}{4})_k(\frac{v}{2}+\frac{5}{4})_k}, \tag{3.24}$$

$$Re\ v > -1, \quad k = 0, 1, 2...,$$

and we have

$$Q_v(z) = \frac{\sqrt{\pi}\Gamma(v+1)}{\Gamma\left(v+\frac{3}{2}\right)(2z)^{v+1}} \sum_{k=0}^{\infty} \frac{\left(\frac{v}{2}+1\right)_k \left(\frac{v}{2}+\frac{1}{2}\right)_k}{\left(v+\frac{3}{2}\right)_k k!} \frac{1}{z^{2k}},$$

$$= \frac{1}{(2z)^{v+1}} \sum_{k=0}^{\infty} \frac{\left(\frac{v}{2}+\frac{3}{4}\right)_k \left(\frac{v}{2}+\frac{5}{4}\right)_k}{\left(v+\frac{3}{2}\right)_k k!} \frac{1}{z^{2k}} \int_1^{\infty} \frac{dt}{t^{2k+v+\frac{3}{2}}\sqrt{t-1}},$$

$$= \frac{1}{(2z)^{v+1}} \int_1^{\infty} F\left(\frac{v}{2}+\frac{3}{4}, \frac{v}{2}+\frac{5}{4}; v+\frac{3}{2}; \frac{1}{z^2 t^2}\right) \frac{dt}{t^{v+\frac{3}{2}}\sqrt{t-1}}.$$
(3.25)

The general solution u of the differential equation (3.2) can be written as a linear combination of Legendre functions of the first and second kinds,

$$u = AQ_v(z) + BP_v(z), \tag{3.26}$$

where |arg(z-1)|<π, $v \neq -1, -2, \ldots$ in the application; it is often necessary to find a general solution of (3.2) for the case where x is a real number in the interval (-1,1). Since $P_v(z)$ is defined for such x, we need to only construct a second linearly independent solution. It is not hard to see that such a solution is given by the function

$$Q_v(x) = \frac{1}{2}[Q_v(x+i0) + Q_v(x-i0)]; \tag{3.27}$$

thus, if $z = x(-1 < x < 1)$, the general solution of (3.2) is

$$u = AP_v(x) + BQ_v(x), \quad v \neq -1, -2... \tag{3.28}$$

3.4 Integral Representations of the Legendre Functions

The Legendre functions have various representations in terms of definite integrals containing the variables z and v as parameters. The most general representations of this type involve contour integrals, but for a practical purpose, representations involving integrals along segments of the real axis are of greatest importance. The integral representations of the function $P_v(z)$ is given by the formula

$$P_v(z) = \frac{2}{\pi} \int_0^{\pi/2} f_v(\frac{z-1}{2}\sin^2\Phi)d\Phi, \qquad |arg(z+1)| < \pi. \quad (3.29)$$

Assuming that $z = \cosh\alpha (\alpha > 0)$ and introducing a new variable of integration in (3.29) by setting

$$\sinh\frac{\theta}{2} = \sinh\frac{\alpha}{2}\sin\Phi,$$

we find that

$$P_v(\cosh\alpha) = \frac{2}{\pi}\int_0^\alpha \frac{\cosh(v+\frac{1}{2})\Theta}{\sqrt{2\cosh\alpha - 2\cosh\Theta}}d\Theta. \quad (3.30)$$

For any real or complex value of the degree v, We write (3.30) in the form

$$P_v(\cosh\alpha) = \frac{1}{\pi}\int_{-\alpha}^\alpha \frac{e^{-(v+\frac{1}{2})\Theta}}{\sqrt{2\cos\alpha - 2\cosh\Theta}}dO,$$

and then setting

$$e^\Theta = \cosh\alpha + \sinh\alpha\cos\psi.$$

The Legendre function of the first kind,

$$P_v(\cosh\alpha) = \frac{1}{\pi}\int_0^\pi \frac{d\psi}{(\cosh\alpha + \sinh\alpha\cos\psi)^{v+1}}, \quad (3.31)$$

where v is arbitrary.

The integral representations of $Q_v(z)$, the Legendre function of the second kind. Assuming that $z = \cosh\alpha(\alpha > 0)$ and introducing a new variable of integration by setting

$$\sinh\frac{\Theta}{2} = \sinh\frac{\alpha}{2}\cosh\psi,$$

we find that

$$Q_v(\cosh\alpha) = \frac{1}{\pi}\int_\alpha^\infty \frac{e^{-(v+\frac{1}{2})\Theta}}{\sqrt{2\cosh\Theta - 2\cosh\alpha}}d\Theta, \qquad (3.32)$$

for Re $v > -1$. Then writing

$$e^\Theta = \cosh\alpha + \sinh\alpha\cos\phi,$$

we reduce (3.32) to the form

$$Q_v(\cosh\alpha) = \int_\alpha^\infty \frac{d\phi}{(2\cosh\phi - \sinh\alpha\cosh\phi)^{v+1}}, \quad \alpha > 0, \ \text{Re}\,v > -1. \qquad (3.33)$$

3.5 Some Relations Satisfied by the Legendre Functions

The differential equation (3.2) does not change if we replace v by $-v-1$ or z by $-z$, and, hence, it has solutions $P_{-v-1}(z), Q_{-v-1}(z), P_v(-z)$, and $Q_v(-z)$, as well as $P_v(z)$ and $Q_v(z)$. Since every three solutions of a second-order linear differential equation are linearly dependent, there must be certain functional relation between the solutions. The simplest such relation is given by the formula

$$P_{-v-1}(z) = P_v(z). \qquad (3.34)$$

To obtain a relation connecting $P_v(z), Q_v(z)$, and $Q_{-v-1}(z)$, we assume temporarily that $z > 1$ and $-1 < \text{Re}\,v < 0$. In this case, $-1 < \text{Re}(-v-1) < 0$, we have

$$Q_v(\cosh\alpha) - Q_{-v-1}(\cosh\alpha) = \pi\cot v\pi P_v(\cosh\alpha),$$

or

$$\sin v\pi\,[Q_v(z) - Q_{-v-1}(z)] = \pi\cot v\pi P_v(z). \qquad (3.35)$$

Formula (3.35) remains valid for all z in the plane cut along $[-\infty, 1]$. Similarly, the given relation for Legendre functions

$$Q_{n-1/2}(z) = Q_{-n-1/2}(z). \qquad (3.36)$$

$$Q_v(-z) = e^{\pm v\pi i} Q_v(z), \qquad v \neq -1, -2, ..., \qquad (3.37)$$

$$\frac{2 \sin v\pi}{\pi} Q_v(z) = P_v(z) e^{\pm v\pi i} - P_v(-z), \qquad (3.38)$$

where $v \neq -1, -2, ...$, and the upper limit is chosen if Im $z > 0$ and the lower sign if im $z < 0$.

The relations (3.35)–(3.38) play an important role in the theory of spherical harmonics. In particular, it follows from (3.38) that

$$\frac{2 \sin v\pi}{\pi} Q_v(x + i0) = P_v(x) e^{-v\pi i} - P_v(-x), \qquad (3.39)$$

$$\frac{2 \sin v\pi}{\pi} Q_v(x - i0) = P_v(x) e^{v\pi i} - P_v(-x), \qquad (3.40)$$

if -1<x<1. This implies that

$$Q_v(x + i0) - Q_v(x - i0) = -i\pi P_v(x), \qquad -1 < x < 1 \quad (3.41)$$

is the Legendre function of the second kind.

3.6 Workskian of Pairs of Solutions of Legendre's Equation

Let $u_1(z)$ and $u_2(z)$ be a pair of solutions of Legendre's equation, with Workskian $W\{u_1(z), u_2(z)\}$. Then

$$\frac{d}{dz}[(1 - z^2) u_1'] + v(v + 1) u_1 = 0,$$

$$\frac{d}{dz}[(1 - z^2) u_2'] + v(v + 1) u_2 = 0,$$

3.6 Workskian of Pairs of Solutions of Legendre's Equation

and subtracting the first equation multiplied by u_2 from the second equation multiplied by u_1, we obtain

$$\frac{d}{dz}[(1-z^2)W\{u_1(z),u_2(z)\}] = 0,$$

which implies

$$W\{u_1(z),u_2(z)\}] = \frac{C}{1-z^2}.$$

In particular, choosing $u_1(z) = Q_v(z)$, $u_2(z) = Q_{-v-1}(z)$, assuming temporarily that $2v$ is not an integer, and letting $|z| \to \infty$, we have

$$u_1(z) = \frac{\sqrt{\pi}\Gamma(v+1)}{\Gamma(v+\frac{3}{2})(2z)^{v+1}}\left[1+O(|z|^{-2})\right],$$

$$u_2(z) = \frac{\sqrt{\pi}\Gamma(-v)}{\Gamma(\frac{1}{2}-v)(2z)^{-1}}\left[1+O(|z|^{-2})\right],$$

$$u_1'(z) = -\frac{\sqrt{\pi}(v+1)\Gamma(v+1)}{\Gamma(v+\frac{3}{2})(2z)^{v+1}z}\left[1+O(|z|^{-2})\right],$$

$$u_2'(z) = \frac{\sqrt{\pi}v\Gamma(-v)}{\Gamma(\frac{1}{2}-v)(2z)^{-v}z}\left[1+O(|z|^{-2})\right].$$

Therefore,

$$W\{u_1(z),u_2(z)\} = \frac{\sqrt{\pi}(v+1)\Gamma(-v)}{2\Gamma(\frac{1}{2}-v)\Gamma(v+\frac{3}{2})}\frac{2v+1}{z^2}\left[1+O(|z|^{-2})\right],$$

$$= -\pi\cot v\pi\frac{1}{z^2}\left[1+O(|z|^{-2})\right].$$

The general solution of Legendre's equation (3.2) can be written in any of the three equivalent forms

$$u = AP_v(z) + BQ_v(z), \qquad |arg(z-1)| < \pi, \quad v \neq -1, -2, ...,$$

$$(3.42)$$

42 Spherical Harmonics Theory

$$u = CP_v(z) + DP_v(-z), \qquad |arg(1 \pm z)| < \pi, \quad v \neq 0, \pm 1, \pm 2, ...,$$
(3.43)

$$u = EQ_v(z) + FQ_{-v-1}(z), \qquad |arg(z-1)| < \pi, \quad 2v \neq 0, \pm 1, \pm 2, ...,$$
(3.44)

where A,B...,F are arbitrary constants.

3.7 Recurrence Relations for the Legendre Functions

The Legendre functions satisfy simple recurrence relations connecting functions with consecutive indices. To derive these relations, we set $z = cosh\alpha$ ($\alpha > 0$), assuming for the time being that z is a real number greater than 1. Then, using the integral representation (3.2), we have

$$P_{v+1}(cosh\alpha) + P_{v-1}(cosh\alpha),$$

$$= \frac{4}{\pi} \int_0^\pi \frac{cosh(v+\frac{1}{2})\Theta cosh\Theta}{\sqrt{2cosh\alpha - 2cosh\Theta}} d\Theta,$$

$$= \frac{4}{\pi} \int_0^\alpha \frac{cosh\alpha cosh(v+\frac{1}{2})\Theta}{\sqrt{2cosh\alpha - 2cosh\Theta}} d\Theta$$

$$- \frac{2}{\pi} \int_0^\alpha \sqrt{2cosh\alpha - 2cosh\Theta} cosh(v+\frac{1}{2})\Theta d\Theta,$$

$$= 2cosh\alpha P_v(cosh\alpha) - \frac{4}{(2v+1)\pi}$$

$$\times \int_0^\alpha \sqrt{2cosh\alpha - 2cosh\Theta} dsinh(v+\frac{1}{2})\Theta,$$

3.7 Recurrence Relations for the Legendre Functions

$$= 2\cosh\alpha P_v(\cosh\alpha) - \frac{4}{(2v+1)\pi}\int_0^\alpha \frac{\sinh(v+\tfrac{1}{2})\Theta \sinh\Theta}{\sqrt{2\cosh\alpha - 2\cosh\Theta}}d\Theta,$$

$$= 2\cosh\alpha P_v(\cosh\alpha) - \frac{2}{(2v+1)\pi}$$

$$\times \int_0^\alpha \frac{\cosh(v+\tfrac{3}{2})\Theta - \cosh(v-\tfrac{1}{2})\Theta}{\sqrt{2\cosh\alpha - 2\cosh\Theta}}d\Theta,$$

$$= 2\cosh\alpha P_v(\cosh\alpha) - \frac{1}{(2v+1)\pi}\left[P_{v+1}(\cosh\alpha) - P_{v-1}(\cosh\alpha)\right],$$

which implies

$$(v+1)P_{v+1}(z) - (2v+1)zP_v(z) + vP_{v-1}(z) = 0. \qquad (3.45)$$

According to the principle of analytic continuation, formula (3.45) holds for arbitrary z in the plane with a cut along the segment $[-\infty, -1]$. In the same way, we find that

$$P_{v+1}(\cosh\alpha) - P_{v-1}(\cosh\alpha) = \frac{4}{\pi}\int_0^\alpha \frac{\sinh(v+\tfrac{1}{2})\Theta \sinh\Theta}{\sqrt{2\cosh\alpha - 2\cosh\Theta}}d\Theta,$$

$$= -\frac{4}{\pi}\int_0^\alpha \sinh(v+\tfrac{1}{2})\Theta\, d\sqrt{2\cosh\alpha - 2\cosh\Theta},$$

$$= (2v+1)\frac{2}{\pi}\int_0^\alpha \sqrt{2\cosh\alpha - 2\cosh\Theta}\cosh(v+\tfrac{1}{2})\Theta\, d\Theta.$$

After differentiation with respect to α, this becomes

$$P'_{v+1}(\cosh\alpha) - P'_{v-1}(\cosh\alpha) = (2v+1)\frac{2}{\pi}\int_0^\alpha \frac{\cosh(v+\tfrac{1}{2})\Theta}{\sqrt{2\cosh\alpha - 2\cosh\Theta}}d\Theta,$$

$$= (2v+1)P_v(\cosh\alpha),$$

or

$$P'_{v+1}(z) - P'_{v-1}(z) = (2v+1)P_v(z), \qquad (3.46)$$

where the result holds in the limit $[-\infty, -1]$. The rest of the recurrence relation is given in the form

$$P'_{v+1}(z) - zP'_v(z) = (v+1)P_v(z). \quad (3.47)$$

$$zP'_v(z) - P'_{v-1}(z) = vP_v(z). \quad (3.48)$$

$$(1-z^2)P'_v(z) = vP_{v-1}(z) - vzP_v(z). \quad (3.49)$$

$$(1-z^2)Q'_v(z) = vQ_{v-1}(z) - vzQ_v(z). \quad (3.50)$$

Formulas (3.46)–(3.50) hold for any complex z in the plane along the limit $[-\infty, -1]$ and for arbitrary $v \neq -1, -2, \ldots$ It is easily verified that these formulas remain valid for the functions $Q_v(x)$.

3.8 Associated Legendre Functions

The next class of spherical harmonics, in order of increasing complexity consists of the associated Legendre functions, which are solutions of the differential equation

$$(1-z^2)u'' - 2zu' + \left[v(v+1) - \frac{m^2}{1-z^2}\right]u = 0, \quad (3.51)$$

for arbitrary v and integral $m = 0, 1, 2, \ldots$ These functions generalize the functions $P_v(z)$ and $Q_v(z)$ and reduce to these functions for $m = 0$. To define the associated Legendre functions, we assume that z is an arbitrary complex number belonging to the plane cut along $[-\infty, 1]$, and we introduce a new function v related to u by the expression

$$u = (z^2 - 1)^{m/2} v = (z-1)^{m/2}(z+1)^{m/2} v.$$

Then (3.51) takes the form

$$(1-z^2)v'' - 2(m+1)zv' + [(v-m)(v+m+1)]v = 0. \quad (3.52)$$

Let w be a solution of Legendre's equation

$$(1-z^2)w'' - 2zw' + v(v+1)w = 0. \tag{3.53}$$

Then it is easily verfied that the function $v = w^{(m)}$ satisfies (3.53). It follows that the solutions of (3.51) are given by

$$P_v^m(z) = (z^2-1)^{m/2}\frac{d^m P_v(z)}{dz^m},$$

$$Q_v^m(z) = (z^2-1)^{m/2}\frac{d^m Q_v(z)}{dz^m}, \quad m=0,1,2..., \tag{3.54}$$

where $P_v(z)$ and $Q_v(z)$ are the Legendre functions. The functions $P_v^m(z)$ and $Q_v^m(z)$ are called the associated Legendre functions of the first and second kinds, respectively. In the application, it is often necessary to find the solution of (3.51) for real $z = x$ belonging to the interval (-1,1). To this end, we first note that values of the associated Legendre functions on the upper and lower edges of the limit are

$$P_v^m(x+i0) = e^{\pm(mni/2)}(1-x^2)^{m/2}\frac{d^m P_v(x)}{dx^m},$$

$$Q_v^m(x+i0) = e^{\pm(mni/2)}(1-x^2)^{m/2}\frac{d^m Q_v(x+i0)}{dx^m}.$$

Then we introduce two new functions $P_v^m(x)$ and $Q_v^m(x)$ by writing

$$P_v^m(x) = e^{+(mni/2)}P_v^m(x+i0) = e^{-(mni/2)}P_v^m(x-i0),$$

$$= (-1)^m(1-x^2)^{m/2}\frac{d^m P_v(x)}{dx^m},$$

$$Q_v^m(x) = \frac{(-1)^m}{2}\left[e^{-(mni/2)}Q_v^m(x+i0) + e^{(mni/2)}Q_v^m(x-i0)\right]$$

$$= (-1)^m(1-x^2)^{m/2}\frac{d^m Q_v(x)}{dx^m},$$

where $-1 < x < 1$, v is arbitrary except that $v \neq -1, -2, ...$

3.9 Exercises

(1) Prove the formulas

$$P_v(-x+i0) - P_v(-x-i0) = 2i\sin v\pi P_v(x),$$

$$Q_v(-x+i0) - Q_v(-x-i0) = 2i\sin v\pi Q_v(x),$$

where x>1.

(2) Derive the following representation of the Legendre function $P_v(z)$ in terms of hypergeometric series:

$$P_v(z) = F\left(\frac{v+1}{2}, -\frac{v}{2}; 1; 1-z^2\right), \quad |1-z^2|, \quad |arg(z+1)| < \pi,$$

$$P_v(z) = \left(\frac{z+1}{2}\right)^v F\left(-v, -v; 1; \frac{z-1}{z+1}\right), \quad \text{Re } z > .$$

(3) Prove that if v is not an integer, then the asymptotic behavior as $z = -1$ of the Legendre function of the first kind and its derivatives is described by the formulas

$$P_v(z) \approx \frac{\sin v\pi}{\pi} \log \frac{z+1}{2},$$

$$P'_v(z) \approx \frac{\sin v\pi}{\pi} \frac{1}{z+1}.$$

(4) Derive the integral representations

$$P_v(\cosh\alpha) = \frac{1}{\Gamma(v+1)} \int_0^\infty e^{-t\cosh\alpha} I_0(t\sinh\alpha) t^v dt,$$

$$Q_v(\cosh\alpha) = \frac{1}{\Gamma(v+1)} \int_0^\infty e^{-t\cosh\alpha} K_0(t\sinh\alpha) t^v dt,$$

$$P_v(\cos\Theta) = \frac{1}{\Gamma(v+1)} \int_0^\infty e^{-t\cos\Theta} J_0(t\sin\Theta) t^v dt,$$

where

$$|Im\alpha| \leqslant \frac{\pi}{2}, \quad \operatorname{Re} v > -1, \quad 0 \leqslant \Theta < \pi,$$

and $J_o(x)$, $I_o(x)$, and $K_o(x)$ are Bessels functions.

(5) Prove the formulas

$$\int_{-1}^{1} P_l^m(x) P_n^m(x) \mathrm{d}x = 0,$$

$$\int_{-1}^{1} [P_n^m(x)]^2 \mathrm{d}x = \frac{2}{2n+1} \frac{(n+m)!}{(n-m)!},$$

$$m = 0, 1, 2, \ldots, l = m, m+1, \ldots, n = m, +1, \ldots$$

(6) Prove that

$$\int_{-1}^{1} x^{2m} P_{2n}(x) \mathrm{d}x = 2^{2n+1} \frac{\Gamma(2m+1)\Gamma(m+n+1)}{\Gamma(m-n+1)\Gamma(2m+2n+2)}.$$

4

Bessel Function

4.1 Bessel Functions

Bessel function is the solution of second-order differential equation that has the form

$$x^2 \frac{d^2y}{dx^2} + x\frac{dy}{dx} + (x^2 - n^2)y = 0 \qquad (4.1)$$

and equation has an essential singularity at $x = \infty$ and n is any integer. Since $x = 0$, is a regular singularity of the equation. We can find the complementary function of the above equation by using the most generalized power series or Frobenius series:

$$y = \sum_{k=0}^{\infty} a_k x^{k+p}. \qquad (4.2)$$

Taking the first and second derivative of (4.2), we get

$$\frac{dy}{dx} = \sum_{k=0}^{\infty} a_k(k+p)x^{k+p-1} \qquad (4.3)$$

$$\frac{d^2y}{dx^2} = \sum_{k=0}^{\infty} a_k(k+p)(k+p-1)x^{k+p-2} \qquad (4.4)$$

50 Bessel Function

Putting (4.2), (4.3), and (4.4) into (4.1), we get

$$\sum_{k=0}^{\infty} a_k(k+p)(k+p-1)x^{k+p} + \sum_{k=0}^{\infty} a_k(k+p)x^{k+p}$$
$$+ (x^2 - n^2)\sum_{k=0}^{\infty} a_k x^{k+p} = 0 \qquad (4.5)$$

Multiplying the above by x^{-p}, we get

$$\sum_{k=0}^{\infty} a_k(k+p)(k+p-1)x^k + \sum_{k=0}^{\infty} a_k(k+p)x^k + (x^2 - n^2)\sum_{k=0}^{\infty} a_k x^k = 0 \qquad (4.6)$$

$$\sum_{k=0}^{\infty} a_k x^k \left((k+p)(k+p-1) + (k+p) + x^2 - n^2\right) = 0 \qquad (4.7)$$

$$\sum_{k=0}^{\infty} a_k x^k \left((k+p)(k+p-1) + (k+p) + x^2 - n^2\right) = 0 \qquad (4.8)$$

$$\sum_{k=0}^{\infty} a_k x^k \left((k+p)(k+p) - n^2\right) + \sum_{k=0}^{\infty} a_k x^{k+2} = 0 \qquad (4.9)$$

Taking x^0 cofficient in (4.9) we obtain,

$$p^2 - n^2 = 0 \qquad (4.10)$$

(4.10) is called Indicial equation and it has two repeated roots, i.e, $p = \pm n$. And for taking x^1 coefficient in eq.(4.9) we obtain,

$$\left((p+1)^2 - n^2\right) a_1 = 0 \qquad (4.11)$$

$$\implies a_1 = 0 \qquad (4.12)$$

Now (4.9) gives

$$\sum_{k=0}^{\infty} \left(a_k x^k \left((k+p)(k+p) - n^2\right) + a_k x^{k+2}\right) = 0 \qquad (4.13)$$

4.1 Bessel Functions

Making (4.13) in x^k by changing $k \to k-2$ in second term, we get

$$\sum_{k=0}^{\infty} x^k \left(a_k \left((k+p)(k+p) - n^2\right) + a_{k-2}\right) = 0 \tag{4.14}$$

$$\sum_{k=0}^{\infty} x^k \left(a_k \left((k+p)^2 - n^2\right) + a_{k-2}\right) = 0 \tag{4.15}$$

Again, (4.15) demands that each cofficients of x^k must vanishes. We get,

$$a_k \left((k+p)^2 - n^2\right) + a_{k-2} = 0 \tag{4.16}$$

Taking root of (4.10) i.e. $p = \pm n$, we get our recurrence relation.

$$a_k = -\frac{1}{k(k \pm 2n)} a_{k-2} \tag{4.17}$$

$k = 2, 3, 4, 5, ...$

Now, take the integer values of k and using (4.12), we find

$$k = 2 \qquad a_2 = -\frac{1}{2(2 \pm 2n)} a_0,$$

$$k = 3, \qquad a_3 = \frac{1}{3(3 \pm 2n)} a_1,$$

$$k = 3, \qquad a_3 = 0$$

$$k = 4, \qquad a_4 = \frac{1}{4(4 \pm 2n)} \frac{1}{2(2 \pm 2n)} a_0,$$

$$k = 5, \qquad a_5 = 0,$$

$$k = 6, \qquad a_6 = -\frac{1}{6(6 \pm 2n)} \frac{1}{4(4 \pm 2n)} \frac{1}{2(2 \pm 2n)} a_0,$$

and so on. Putting value of $a_0, a_1, a_2, a_3, ...$ in (4.2), we get

$$y = a_0 \left(1 - \frac{1}{2(2 \pm 2n)} x^{2 \pm n} + \frac{1}{4.(4 \pm 2n)} \frac{1}{2.(2 \pm 2n)} x^{4 \pm n}\right.$$

52 Bessel Function

$$-\frac{1}{6(6\pm 2n)}\frac{1}{4(4\pm 2n)}\frac{1}{2(2\pm 2n)}x^{6\pm n}+\cdots\Big) \qquad (4.18)$$

$$y = a_0 x^{\pm n}\left(1 - \frac{1}{2(2\pm 2n)}x^2 + \frac{1}{4.(4\pm 2n)}\frac{1}{2.(2\pm 2n)}x^4\right.$$

$$\left.-\frac{1}{6(6\pm 2n)}\frac{1}{4(4\pm 2n)}\frac{1}{2(2\pm 2n)}x^6+\cdots\right) \qquad (4.19)$$

$$y = a_0 x^{\pm n}\left(1 - \frac{1}{(1\pm n)}\left(\frac{x}{2}\right)^2 + \frac{1}{2!(2\pm n)(1\pm n)}\left(\frac{x}{2}\right)^4\right.$$

$$\left.-\frac{1}{3!(3\pm n)(2\pm n)(1\pm n)}\left(\frac{x}{2}\right)^6+\cdots\right) \qquad (4.20)$$

Now, choosing the value of $a_0 = \frac{1}{2^{\pm n}\Gamma(1\pm n)}$, we get the Bessel function denoted by $J_n(x)$

$$J_{\pm n}(x) = \frac{1}{2^{\pm n}\Gamma(1\pm n)}x^{\pm n}(1 - \frac{1}{(1\pm n)}\left(\frac{x}{2}\right)^2 + \frac{1}{2!(2\pm n)(1\pm n)}\left(\frac{x}{2}\right)^4 -$$

$$-\frac{1}{3!(3\pm n)(2\pm n)(1\pm n)}\left(\frac{x}{2}\right)^0+\cdots) \qquad (4.21)$$

$$J_{\pm n}(x) = \frac{1}{\Gamma(1\pm n)}\left(\frac{x}{2}\right)^{\pm n}(1 - \frac{1}{(1\pm n)}\left(\frac{x}{2}\right)^2 + \frac{1}{2!(2\pm n)(1\pm n)}\left(\frac{x}{2}\right)^4 -$$

$$-\frac{1}{3!(3\pm n)(2\pm n)(1\pm n)}\left(\frac{x}{2}\right)^6+\cdots) \qquad (4.22)$$

$$J_{\pm n}(x) = \sum_{m=0}^{\infty}\frac{(-1)^m}{m!\,\Gamma(1\pm n+m)}\left(\frac{x}{2}\right)^{\pm n+2m} \qquad (4.23)$$

where Γn is a Gamma function and it is given by $\Gamma n = n(n-1)!$.

4.1 Bessel Functions

The general solution of eq.(4.1) has the form,

$$y = AJ_n(x) + BJ_{-n}(x) \tag{4.24}$$

where, $J_n(x)$ and $J_{-n}(x)$ are called Bessel function of first kind and it is governed by (4.23).

Since Bessel polynomials $(J_{\pm n}(x))$ are solutions of (4.1). Hence, it satisfies the (4.1), and we get

$$x^2 J''_{\pm n}(x) + x J'_{\pm n}(x) + (x^2 - n^2) J_{\pm n}(x) = 0. \tag{4.25}$$

For examples, for $n = 1/2$, (4.1) becomes

$$x^2 \frac{d^2y}{dx^2} + x\frac{dy}{dx} + (x^2 - \frac{1}{4})y = 0 \tag{4.26}$$

The solution of (4.26) is found by putting the value of $n = 1/2$ in (4.23) and we get

$$J_{\pm 1/2}(x) = \sum_{m=0}^{\infty} \frac{(-1)^m}{m!\,\Gamma(\pm 1/2 + m + 1)} \left(\frac{x}{2}\right)^{\pm 1/2 + 2m} \tag{4.27}$$

(4.27) contains two series. The series form of Bessel function is found by (4.27) using the relation, i.e., $\Gamma(m+1) = m\Gamma m$; and $\Gamma 1/2 = \sqrt{\pi}$

$$J_{1/2}(x) = \sum_{m=0}^{\infty} \frac{(-1)^m}{m!\,\Gamma(1/2 + m + 1)} \left(\frac{x}{2}\right)^{1/2 + 2m}$$

$$J_{1/2}(x) = \sum_{m=0}^{\infty} \frac{(-1)^m}{m!\,\Gamma(m + \frac{3}{2})} \left(\frac{x}{2}\right)^{1/2 + 2m}$$

$$J_{1/2}(x) = \frac{x^{1/2}}{2^{1/2}\Gamma\frac{3}{2}} - \frac{x^{5/2}}{1!2^{5/2}\Gamma\frac{5}{2}} + \frac{x^{9/2}}{2!2^{9/2}\Gamma\frac{7}{2}} - \frac{x^{13/2}}{3!2^{13/2}\Gamma\frac{9}{2}} + \dots$$

$$J_{1/2}(x) = \frac{x^{1/2}}{2^{1/2}\frac{1}{2}\Gamma\frac{1}{2}} - \frac{x^{5/2}}{1!2^{5/2}\frac{3}{2}\frac{1}{2}\Gamma\frac{1}{2}} + \frac{x^{9/2}}{2!2^{9/2}\frac{5}{2}\frac{3}{2}\frac{1}{2}\Gamma\frac{1}{2}} - \frac{x^{13/2}}{3!2^{13/2}\frac{7}{2}\frac{5}{2}\frac{3}{2}\frac{1}{2}\Gamma\frac{1}{2}} + \dots$$

$$J_{1/2}(x) = \frac{x^{1/2}}{2^{1/2}\frac{1}{2}\sqrt{\pi}} - \frac{x^{5/2}}{2^{5/2}\frac{3}{2}\frac{1}{2}\sqrt{\pi}} + \frac{x^{9/2}}{2 \cdot 2^{9/2}\frac{5}{2}\frac{3}{2}\frac{1}{2}\sqrt{\pi}} - \frac{x^{13/2}}{3 \cdot 2 \cdot 2^{13/2}\frac{7}{2}\frac{5}{2}\frac{3}{2}\frac{1}{2}\sqrt{\pi}} + \ldots$$

$$J_{1/2}(x) = \frac{x^{1/2}}{2^{1/2}\frac{1}{2}\sqrt{\pi}}\left(1 - \frac{x^2}{2^2\frac{3}{2}} + \frac{x^4}{2 \cdot 2^4\frac{5}{2}\frac{3}{2}} - \frac{x^6}{3 \cdot 2 \cdot 2^6\frac{7}{2}\frac{5}{2}\frac{3}{2}} + \ldots\right)$$

$$J_{1/2}(x) = \frac{x^{1/2}}{2^{1/2}\frac{1}{2}\sqrt{\pi}}\left(1 - \frac{x^2}{3.2} + \frac{x^4}{5.4.3.2} - \frac{x^6}{7.6.5.4.3.2.} + \ldots\right)$$

$$J_{1/2}(x) = \frac{x^{1/2}}{2^{1/2}\frac{1}{2}\sqrt{\pi}}\left(1 - \frac{x^2}{3!} + \frac{x^4}{5!} - \frac{x^6}{7!.} + \ldots\right)$$

$$J_{1/2}(x) = \frac{x^{1/2}}{2^{1/2}\frac{1}{2}\sqrt{\pi}x}\left(x - \frac{x^3}{3!} + \frac{x^5}{5!} - \frac{x^7}{7!.} + \ldots\right)$$

$$J_{1/2}(x) = \frac{\sin x}{x^{1/2}\frac{1}{2^{1/2}}\sqrt{\pi}}$$

$$J_{1/2}(x) = \sqrt{\frac{2}{\pi x}}\sin x. \tag{4.28}$$

Similarly, for $n = -1/2$,

$$J_{-1/2}(x) = \sqrt{\frac{2}{\pi x}}\cos x. \tag{4.29}$$

4.2 Generating Function

The generating function of Bessel function $J_n(x)$ is defined as

$$g(x,t) = e^{\frac{x}{2}(t-\frac{1}{t})} = \sum_{n=0}^{\infty} J_n(x)t^n \tag{4.30}$$

We know that,

$$e^x = \sum_{n=0}^{\infty}\frac{x^n}{n!} = 1 + \frac{x}{1!} + \frac{x^2}{2!} + \frac{x^3}{3!} + \ldots$$

4.2 Generating Function

Now, the left-hand side of (4.30) has the form

$$e^{\frac{x}{2}t} \cdot e^{-\frac{x}{2t}} = \sum_{n=0}^{\infty} \frac{\left(\frac{xt}{2}\right)^n}{n!} \cdot \frac{\left(\frac{-x}{2t}\right)^n}{n!}$$

$$e^{\frac{x}{2}t} = 1 + \frac{\left(\frac{x}{2}t\right)}{1!} + \frac{\left(\frac{x}{2}t\right)^2}{2!} + \frac{\left(\frac{x}{2}t\right)^3}{3!} + \ldots + \frac{\left(\frac{x}{2}t\right)^n}{n!} + \frac{\left(\frac{x}{2}t\right)^{n+1}}{(n+1)!} + \frac{\left(\frac{x}{2}t\right)^{n+2}}{(n+2)!} + \ldots \quad (4.31)$$

$$e^{-\frac{x}{2t}} = 1 - \frac{\left(\frac{x}{2t}\right)}{1!} + \frac{\left(\frac{x}{2t}\right)^2}{2!} - \frac{\left(\frac{x}{2t}\right)^3}{3!} + \ldots + \frac{\left(\frac{x}{2t}\right)^n}{n!} - \frac{\left(\frac{x}{2t}\right)^{n+1}}{(n+1)!} + \frac{\left(\frac{x}{2t}\right)^{n+2}}{(n+2)!} - \ldots \quad (4.32)$$

Multiplying (4.31) and (4.32), we get

$$\Longrightarrow \left(1 + \frac{\left(\frac{x}{2}t\right)}{1!} + \frac{\left(\frac{x}{2}t\right)^2}{2!} + \frac{\left(\frac{x}{2}t\right)^3}{3!} + \ldots + \frac{\left(\frac{x}{2}t\right)^n}{n!} + \frac{\left(\frac{x}{2}t\right)^{n+1}}{(n+1)!} + \frac{\left(\frac{x}{2}t\right)^{n+2}}{(n+2)!} + \ldots\right) \times$$

$$\left(1 - \frac{\left(\frac{x}{2t}\right)}{1!} + \frac{\left(\frac{x}{2t}\right)^2}{2!} - \frac{\left(\frac{x}{2t}\right)^3}{3!} + \ldots + \frac{\left(\frac{x}{2t}\right)^n}{n!} - \frac{\left(\frac{x}{2t}\right)^{n+1}}{(n+1)!} + \frac{\left(\frac{x}{2t}\right)^{n+2}}{(n+2)!} - \ldots\right) \quad (4.33)$$

$$\Longrightarrow \ldots \frac{\left(\frac{x}{2}\right)^n t^n}{n!} - \frac{\left(\frac{x}{2}\right)^{n+1} t^{n-1}}{1! n!} + \frac{\left(\frac{x}{2}\right)^{n+2} t^{n-2}}{2! n!} - \ldots + \frac{\left(\frac{x}{2}\right)^{n+1} t^{n+1}}{(n+1)!}$$

$$- \frac{\left(\frac{x}{2}\right)^{n+2} t^n}{1!(n+1)!} + \frac{\left(\frac{x}{2}\right)^{n+3} t^{n-1}}{2!(n+1)!} - \ldots$$

$$\ldots + \frac{\left(\frac{x}{2}\right)^{n+2} t^{n+2}}{(n+2)!} - \frac{\left(\frac{x}{2}\right)^{n+1} t^{n+1}}{1!(n+2)!} + \frac{\left(\frac{x}{2}\right)^{n+4} t^n}{2!(n+2)!} - \frac{\left(\frac{x}{2}\right)^{n+5} t^{n-1}}{3!(n+2)!} + \ldots \quad (4.34)$$

Collecting the coefficient of t^n from (4.34), we get

$$\Longrightarrow \frac{\left(\frac{x}{2}\right)^n}{n!} - \frac{\left(\frac{x}{2}\right)^{n+2}}{1!(n+1)!} + \frac{\left(\frac{x}{2}\right)^{n+4}}{2!(n+2)!} - \ldots$$

$$\Longrightarrow \sum_{r=0}^{\infty} \frac{(-1)^r \left(\frac{x}{2}\right)^{n+2r}}{r!(n+r)!}$$

56 Bessel Function

using the fact $m! = \Gamma(m+1)$,

$$\Longrightarrow \sum_{r=0}^{\infty} \frac{(-1)^r (\frac{x}{2})^{n+2r}}{r!\Gamma(n+r+1)} \tag{4.35}$$

from (4.30) and (4.35), we can write

$$J_n(x) = \sum_{r=0}^{\infty} \frac{(-1)^r (\frac{x}{2})^{n+2r}}{r!\Gamma(n+r+1)} \tag{4.36}$$

Now change the sign of n by $-n$,

$$J_{-n}(x) = \sum_{r=0}^{\infty} \frac{(-1)^r (\frac{x}{2})^{-n+2r}}{r!\Gamma(-n+r+1)}$$

and replace variable, i.e., $r \to n+p$,

$$J_{-n}(x) = \sum_{r=0}^{\infty} \frac{(-1)^{n+p} (\frac{x}{2})^{n+2p}}{(n+p)!\Gamma(p+1)}$$

$$J_{-n}(x) = (-1)^n \sum_{p=0}^{\infty} \frac{(-1)^p (\frac{x}{2})^{n+2p}}{p!\Gamma(n+p+1)}$$

$$\boldsymbol{J_{-n}(x) = (-1)^n J_n(x)}. \tag{4.37}$$

Now, putting $n = 0, 1, \ldots$ in (4.36), we get

$$J_0(x) = \sum_{r=0}^{\infty} \frac{(-1)^r (\frac{x}{2})^{2r}}{r!\Gamma(r+1)}$$

$$J_0(x) = 1 - \frac{(\frac{x}{2})^2}{1!} + \frac{(\frac{x}{2})^4}{2!2!} - \frac{(\frac{x}{2})^6}{3!3!} + \ldots$$

and

$$J_1(x) = \sum_{r=0}^{\infty} \frac{(-1)^r (\frac{x}{2})^{1+2r}}{r!\Gamma(r+2)}$$

$$J_1(x) = \frac{\frac{x}{2}}{1!} - \frac{(\frac{x}{2})^3}{2!} + \frac{(\frac{x}{2})^5}{2!3!} - \frac{(\frac{x}{2})^7}{3!4!} + \ldots$$

4.3 Recurrence Relations

We know that
$$J_n(x) = \sum_{r=0}^{\infty} \frac{(-1)^r (\frac{x}{2})^{n+2r}}{r!\Gamma(n+r+1)} \tag{4.38}$$

multiplying the above equation by x^n,

$$x^n J_n(x) = x^n \sum_{r=0}^{\infty} \frac{(-1)^r (\frac{x}{2})^{n+2r}}{r!\Gamma(n+r+1)}$$

$$x^n J_n(x) = \sum_{r=0}^{\infty} \frac{(-1)^r (\frac{1}{2})^{n+2r} x^{2n+2r}}{r!\Gamma(n+r+1)} \tag{4.39}$$

Differentiating (4.39) w.r.t. x, we get

$$\frac{d}{dx}(x^n J_n(x)) = \frac{d}{dx}\left(\sum_{r=0}^{\infty} \frac{(-1)^r (\frac{1}{2})^{n+2r} x^{2n+2r}}{r!\Gamma(n+r+1)}\right)$$

$$\frac{d}{dx}(x^n J_n(x)) = \sum_{r=0}^{\infty} \frac{(2n+2r)(-1)^r (\frac{1}{2})^{n+2r} x^{2n+2r-1}}{r!\Gamma(n+r+1)}$$

$$\frac{d}{dx}(x^n J_n(x)) = x^n \sum_{r=0}^{\infty} \frac{(2n+2r)(-1)^r (\frac{1}{2})^{n+2r} x^{n+2r-1}}{r!(n+r)!}$$

$$\frac{d}{dx}(x^n J_n(x)) = x^n \sum_{r=0}^{\infty} \frac{(n+r)(-1)^r (\frac{1}{2})^{n-1+2r} x^{n-1+2r}}{r!(n+r)(n+r-1)!}$$

$$\frac{d}{dx}(x^n J_n(x)) = x^n \sum_{r=0}^{\infty} \frac{(-1)^r (\frac{x}{2})^{(n-1)+2r}}{r!((n-1)+r)!}$$

Bessel Function

$$\frac{d}{dx}(x^n J_n(x)) = x^n \sum_{r=0}^{\infty} \frac{(-1)^r (\frac{x}{2})^{(n-1)+2r}}{r! \Gamma((n-1)+r+1)}$$

$$\frac{d}{dx}(x^n J_n(x)) = x^n J_{n-1}(x). \qquad (4.40)$$

Similarly,

$$x^{-n} J_n(x) = \sum_{r=0}^{\infty} \frac{(-1)^r (\frac{1}{2})^{n+2r} x^{2r}}{r! \Gamma(-n+r+1)}$$

$$\frac{d}{dx}\left(x^{-n} J_n(x)\right) = \frac{d}{dx}\left(\sum_{r=0}^{\infty} \frac{(-1)^r (\frac{1}{2})^{n+2r} x^{2r}}{r! \Gamma(n+r+1)}\right)$$

$$\frac{d}{dx}\left(x^{-n} J_n(x)\right) = \sum_{r=0}^{\infty} \frac{(2r)(-1)^r (\frac{1}{2})^{n+2r} x^{2r-1}}{r(r-1)! \Gamma(n+r+1)}$$

$$\frac{d}{dx}\left(x^{-n} J_n(x)\right) = x^n \sum_{r=0}^{\infty} \frac{(2)(-1)^r (\frac{1}{2})^{n+2r} x^{2r-1}}{(r-1)! \Gamma(n+r+1) x^n}$$

$$\frac{d}{dx}\left(x^{-n} J_n(x)\right) = x^{-n} \sum_{r=0}^{\infty} \frac{(-1)^r (\frac{1}{2})^{n+2r-1} x^{2r+n-1}}{(r-1)! \Gamma(n+r+1)}$$

change $r \to p+1$

$$\frac{d}{dx}\left(x^{-n} J_n(x)\right) = x^{-n} \sum_{r=0}^{\infty} \frac{(-1)^{p+1} (\frac{1}{2})^{n+2p+1} x^{2p+n+1}}{(p)! \Gamma(n+p+2)}$$

$$\frac{d}{dx}(x^n J_n(x)) = -x^{-n} \sum_{r=0}^{\infty} \frac{(-1)^p (\frac{x}{2})^{(n+1)+2p}}{p! \Gamma((n+1)+p+1)}$$

$$\frac{d}{dx}\left(x^{-n} J_n(x)\right) = -x^n J_{n+1}(x). \qquad (4.41)$$

Now, multiplying (4.40) by x^{-n}, we get

$$x^{-n}\frac{\mathrm{d}}{\mathrm{d}x}(x^n J_n(x)) = J_{n-1}(x)$$

$$x^{-n}\left\{x^n J_n'(x) + J_n(x)nx^{n-1}\right\} = J_{n-1}(x)$$

$$x^{-n}x^n J_n'(x) + x^{-n} J_n(x)nx^{n-1} = J_{n-1}(x)$$

$$J_n'(x) + \frac{n}{x}J_n(x) = J_{n-1}(x). \qquad (4.42)$$

Similarly, multiplying (4.41) by x^{-n}, we get

$$x^{-n}\frac{\mathrm{d}}{\mathrm{d}x}(x^{-n} J_n(x)) = -J_{n+1}(x)$$

$$x^{-n}\left\{x^{-n} J_n'(x) + J_n(x)(-n)x^{n-1}\right\} = -J_{n+1}(x)$$

$$x^{-n}x^n J_n'(x) - x^{-n} J_n(x)nx^{n-1} = -J_{n+1}(x)$$

$$J_n'(x) - \frac{n}{x}J_n(x) = -J_{n+1}(x). \qquad (4.43)$$

4.4 Orthonormality

The orthogonality of Bessel function is defined by a multiplicative weight factor $u = x$, i.e.,

$$\int_0^1 x J_n(\alpha x) J_n(\beta x) \mathrm{d}x = 0 \qquad if \; \alpha \neq \beta \qquad (4.44)$$

60 Bessel Function

and the normality is defined as,

$$\int_0^1 x J_n(\alpha x) J_n(\beta x) \mathrm{d}x = \frac{1}{2} J_{n+1}^2(\alpha) \qquad if\ \alpha = \beta.$$
(4.45)

4.5 Application to the Optical Fiber

Bessel function is frequently comes into the solution of problem possesses cylindrical symmetry (r, ϕ, z) like fiber optical cable. Optical fiber is based on the principle of total internal reflections. An optical fiber has a core of refractive index n_1 and cladding of refractive index n_2. Core refractive index is always greater than the cladding refractive index. This refractive index variation decides the type of optical fibers. Here, for simplicity, we are considering the step index optical fiber whose refractive indexes are constant, i.e.,

$$n(r) = \left\{ \begin{array}{lll} n_1 & 0 < r < s & core \\ n_2 & r > s & cladding \end{array} \right\}.$$
(4.46)

Since actual fiber has a cylindrical structure. Hence, we are using the cylindrical coordinate system, i.e., r, ϕ, z and

We know the scalar wave equation

$$\nabla^2 \psi(r, \phi, z, t) - \frac{n^2}{c^2} \frac{\partial^2 \psi(r, \phi, z, t)}{\partial t^2} = 0$$
(4.47)

where,

$$c = \frac{1}{\sqrt{\epsilon_0 \mu_0}}$$

The solution of (4.47) has the form

$$\psi(r, \phi, z, t) = \psi(r, \phi) e^{i(\omega t - \beta z)}.$$
(4.48)

4.5 Application to the Optical Fiber

The schrodinger equation in cylindrical coordinates form is given by,

$$\nabla^2 \psi(r, \phi, z) + \frac{2m}{\hbar^2}(E - V(r))\psi(r, \phi, z) = 0 \quad (4.49)$$

where,

$$\nabla^2 \psi = \frac{\partial^2 \psi}{\partial r^2} + \frac{1}{r}\frac{\partial \psi}{\partial r} + \frac{1}{r^2}\frac{\partial^2 \psi}{\partial \phi^2} + \frac{\partial^2 \psi}{\partial z^2}. \quad (4.50)$$

From (4.48), we get

$$\frac{\partial^2 \psi}{\partial t^2} = -\omega^2 \psi \quad (4.51)$$

and

$$\frac{\partial^2 \psi}{\partial z^2} = -\beta^2 \psi. \quad (4.52)$$

(4.58) must satisfy (4.47), and using (4.50), (4.51), and (4.52), we get

$$\frac{\partial^2 \psi}{\partial r^2} + \frac{1}{r}\frac{\partial \psi}{\partial r} + \frac{1}{r^2}\frac{\partial^2 \psi}{\partial \phi^2} + (k^2 n^2 - \beta^2)\psi(r, \phi) = 0 \quad (4.53)$$

where

$$k = \frac{\omega}{c} = \frac{2\pi}{\lambda}.$$

Now using the method of separation of variable,

$$\psi(r, \phi) = R(r)\Phi(\phi). \quad (4.54)$$

Putting (4.54) into (4.53) and multiplying by $\frac{r^2}{\psi(r,\phi)}$, we get

$$\frac{r^2}{R}\left(\frac{d^2 R}{dr^2} + \frac{1}{r}\frac{dR}{dr}\right) + r^2(k^2 n^2 - \beta^2) = -\frac{1}{\Phi}\frac{\partial^2 \Phi}{\partial \phi^2} = p^2. \quad (4.55)$$

We are considering only the radial part, i.e.,

$$\frac{r^2}{R}\left(\frac{d^2 R}{dr^2} + \frac{1}{r}\frac{dR}{dr}\right) + r^2(k^2 n^2 - \beta^2) = p^2 \quad (4.56)$$

$$r^2 \frac{d^2 R}{dr^2} + r\frac{dR}{dr} + \left(r^2(k^2 n^2 - \beta^2) - p^2\right) R = 0. \quad (4.57)$$

Bessel Function

For the refractive index defined in (4.46), (4.57) becomes,

$$r^2 \frac{d^2R}{dr^2} + r\frac{dR}{dr} + \left(r^2(k^2 n_1^2 - \beta^2) - p^2\right) R = 0 \qquad 0 < r < s \tag{4.58}$$

$$r^2 \frac{d^2R}{dr^2} + r\frac{dR}{dr} + \left(X^2 \frac{r^2}{s^2} - p^2\right) R = 0, \qquad X = s\sqrt{k^2 n_1^2 - \beta^2} \tag{4.59}$$

and

$$r^2 \frac{d^2R}{dr^2} + r\frac{dR}{dr} + \left(r^2(k^2 n_2^2 - \beta^2) - p^2\right) R = 0 \qquad r > s \tag{4.60}$$

$$r^2 \frac{d^2R}{dr^2} + r\frac{dR}{dr} - \left(Y^2 \frac{r^2}{s^2} - p^2\right) R = 0, \qquad Y = s\sqrt{\beta^2 - k^2 n_2^2} \tag{4.61}$$

Now, we define a very important quantity called V-parameter, i.e.,

$$V^2 = X^2 + Y^2$$

$$V = ks\sqrt{n_1^2 - n_2^2} = \frac{2\pi}{\lambda} s\sqrt{n_1^2 - n_2^2}. \tag{4.62}$$

The solution of (4.59) and (4.61) are,

$$\psi(r, \phi) = \begin{pmatrix} \frac{A}{J_p(X)} J_p(\frac{Xr}{s}) & r < s \\ \frac{A}{J_p(Y)} J_p(\frac{Yr}{s}) & r > s \end{pmatrix}.$$

Here, $J_p(\frac{Xr}{s})$ and $J_p(\frac{Yr}{s})$ are the modified Bessel function.

4.6 Exercises

1. Show that

    a) $2nJ_n(x) = x\left(J_{n-1}(x) + J_{n+1}(x)\right)$

    b) $J(x) = -J_1(x)$

c) $J(x) + J_{n+5}(x) = \dfrac{2}{x}(n+4)J_{n+4}(x)$

d) $\dfrac{\mathrm{d}}{\mathrm{d}x}\left(xJ_n(x)J_{n+1}(x)\right) = x\left(J_n^2(x) - J_{n+1}^2(x)\right).$

2. Prove the orthogonality of the Bessel function given in eq.(4.44).
3. Find the value of $J_{3/2}(x)$, $J_{-3/2}(x)$, $J_{5/2}(x)$, $and\ J_{-5/2}(x)$.
4. Express $J_5(x)$ in terms of $J_0(x)$ and $J_1(x)$.

5

Hermite Polynomials

5.1 Hermite Functions

Hermite function is the solution of second-order differential equation has the form

$$\frac{d^2y}{dx^2} - 2x\frac{dy}{dx} + 2ny = 0, \tag{5.1}$$

and the equation has an essential singularity at $x = \infty$ and n is an any integer. Since $x = 0$ is an ordinary point of the equation, we can find the complementary function of the above equation by using the power series:

$$y = \sum_{k=0}^{\infty} a_k x^k. \tag{5.2}$$

Taking first and second derivative of (5.2), we get

$$\frac{dy}{dx} = \sum_{k=0}^{\infty} a_k k x^{k-1} \tag{5.3}$$

$$\frac{d^2y}{dx^2} = \sum_{k=0}^{\infty} a_k k(k-1) x^{k-2}. \tag{5.4}$$

Putting (5.2), (5.3), and (5.4) into (5.1), we get

$$\sum_{k=0}^{\infty} a_k k(k-1) x^{k-2} - 2 \sum_{k=0}^{\infty} a_k k x^k + 2n \sum_{k=0}^{\infty} a_k x^k = 0. \tag{5.5}$$

Hermite Polynomials

Making (5.5) in x^k by changing $k \to k+2$, we get

$$\sum_{k=0}^{\infty} x^k \left(a_{k+2}(k+1)(k+2) - 2a_k k + 2na_k\right) = 0 \qquad (5.6)$$

$$\sum_{k=0}^{\infty} x^k \left(a_{k+2}(k+1)(k+2) + (-2k + 2n)a_k\right) = 0 \qquad (5.7)$$

$$\sum_{k=0}^{\infty} x^k \left(a_{k+2}(k+1)(k+2) + (2n - 2k)a_k\right) = 0. \qquad (5.8)$$

(5.8) demands that each power of x must vanish, and we obatin

$$a_{k+2}(k+1)(k+2) + (2n - 2k)a_k = 0$$

$$a_{k+2} = \frac{2(k-n)}{(k+1)(k+2)} a_k \qquad (5.9)$$

$k = 0, 1, 2, 3, 4, 5, \ldots$

(5.9) is called recurrence relation. Now, take the integer values of k,

$$k = 0, \qquad a_2 = \frac{2(-n)}{2.1} a_0,$$

$$k = 1, \qquad a_3 = \frac{2(1-n)}{2.3} a_1,$$

$$k = 2 \qquad a_4 = \frac{2(2-n)}{3.4.2} \cdot \frac{2(-n)}{2} a_0,$$

$$k = 3, \qquad a_5 = \frac{2(3-n)}{5.4} \cdot \frac{2(1-n)}{3.2} a_1,$$

$$k = 4, \qquad a_6 = \frac{2(4-n)}{5.6} \cdot \frac{2(2-n)}{3.4.2} \cdot \frac{2(-n)}{2} a_0,$$

5.1 Hermite Functions

$$k = 5, \quad a_7 = \frac{2(5-n)}{7 \cdot 6} \cdot \frac{2(3-n)}{5 \cdot 4} \cdot \frac{2(1-n)}{3 \cdot 2} a_1,$$

and so on. The explicit form of eq.(5.2) can be written as,

$$y = a_0 + a_1 x + a_2 x^2 + a_3 x^3 + a_4 x^4 + a_5 x^5 + a_6 x^6 + a_7 x^7 + \ldots \tag{5.10}$$

Putting the value of $a_0, a_1, a_2, a_3, \ldots$ in (5.10), we get,

$$y = a_0 \left(1 - \frac{2n}{2!} x^2 - \frac{2^2(2-n)}{4!} n.x^4 - \frac{2^3(4-n)}{6!}(2-n)n.x^6 + \ldots \right)$$

$$+ a_1 \left(x + \frac{2(1-n)}{3!} x^3 + \frac{2^2(3-n)}{5!}(1-n)x^5 \right.$$

$$\left. + \frac{2^3(5-n)}{7!}(3-n)(1-n)x^7 + \ldots \right). \tag{5.11}$$

(5.11) is our solution of second-order differential (5.1). Now, taking the integer values of n and varying the choice of $a_0\ and\ a_1$ our solution y becomes a polynomial termed as Hermite polynomials.

$$n = 0, \quad y = a_0,$$

$$n = 1, \quad y = a_1 x,$$

$$n = 2 \quad y = a_0(1 - 2x^2),$$

$$n = 3, \quad y = a_1\left(x - \frac{2}{3}x^3\right),$$

$$n = 4, \quad y = a_0\left(1 - 4x^2 + \frac{2^3}{6}x^4\right),$$

$$n = 5, \quad y = a_1\left(x - \frac{4}{3}x^3 + \frac{2^2}{15}x^5\right).$$

Now, by tuning the value of a_0 and a_1, i.e.,

$$a_0 = 1, \quad H_0(x) = 1,$$

$$a_1 = 2, \quad H_1(x) = 2x,$$

$$a_0 = -2 \quad H_2(x) = (2^2 x^2 - 2),$$

$$a_1 = -12, \quad H_3(x) = (2^3 x^3 - 12x),$$

$$a_0 = 12, \quad H_4(x) = (2^4 x^4 - 48x^2 + 12),$$

$$a_1 = 120, \quad H_5(x) = (2^5 x^5 - 160x^3 + 120x)$$

and so on, you can find the Hermite polynomials. Since Hermite polynomials $(H_n(x))$ are solutions of (5.1), it satisfies (5.1), we get

$$H_n''(x) - 2x H_n'(x) + 2n H_n(x) = 0. \tag{5.12}$$

The general expression for Hermite polynomial is given by

$$H_n(x) = \sum_{m=0}^{[n/2]} (-1)^m \frac{n!}{n!(n-2m)!} (2x)^{n-2m}$$

where $[n/2]$ denotes the integer part of $n/2$, i.e.,

$$[n/2] = \begin{pmatrix} n/2 & if\ n\ is\ even \\ (n-1)/2 & if\ n\ is\ odd \end{pmatrix}.$$

We can generates the hermite polynomial using the above expression by terminating series for integral n.

Hermite functions for even values of n are even functions and odd values of n are odd functions. Since Hermite functions are the solutions of second-order differential equation. Hence, in Physics problems, if you tackle with some symmetric or periodic potentials in second ODE, then naturally the solution decomposes into even and odd functions like the solution of Harmonic oscillator potential in quantum mechanics.

5.2 Generating Function

The generating function of Hermite function $H_n(x)$ is define as

$$g(x,t) = e^{-t^2+2tx} = \sum_{n=0}^{\infty} H_n(x) \frac{t^n}{n!}. \tag{5.13}$$

We know that

$$e^x = \sum_{n=0}^{\infty} \frac{x^n}{n!} = 1 + \frac{x}{1!} + \frac{x^2}{2!} + \frac{x^3}{3!} + \dots$$

Now, the left-hand side of (5.13) has the form

$$e^{-t^2+2tx} = \sum_{n=0}^{\infty} \frac{(-t^2+2xt)^n}{n!}$$

$$\sum_{n=0}^{\infty} \frac{(-t^2+2xt)^n}{n!} = 1 + \frac{(-t^2+2xt)}{1!} + \frac{(-t^2+2xt)^2}{2!} + \frac{(-t^2+2xt)^3}{3!} + \dots$$

Putting the above series into (5.13), we get

$$1 + \frac{(-t^2+2xt)}{1!} + \frac{(-t^2+2xt)^2}{2!} + \frac{(-t^2+2xt)^3}{3!} + \dots = \sum_{n=0}^{\infty} H_n(x) \frac{t^n}{n!}$$

$$1 + \frac{(-t^2+2xt)}{1!} + \frac{(-t^2+2xt)^2}{2!} + \dots = H_0(x) + H_1(x)\frac{t}{1!} + H_2(x)\frac{t^2}{2!} + \dots$$

$$1 + (2xt - t^2) + \frac{t^4 + 4t^2x^2 - 4t^3x}{2!} + \dots = H_0(x) + H_1(x)\frac{t}{1!} + H_2(x)\frac{t^2}{2!} + \dots$$

$$1 + 2xt + t^2\frac{(4x^2 - 2)}{2!} + \dots = H_0(x) + H_1(x)\frac{t}{1!} + H_2(x)\frac{t^2}{2!} + \dots$$

Table 5.1 Some Hermite Polynomials

$H_0(x) =$ 1
$H_1(x) =$ $2x$
$H_2(x) =$ $2^2 x^2 - 2$
$H_3(x) =$ $2^3 x^3 - 12x$
$H_4(x) =$ $2^4 x^4 - 48x^2 + 12$
$H_5(x) =$ $2^5 x^5 - 160x^3 + 120x$
$H_6(x) =$ $2^6 x^6 - 480x^4 + 720x^2 - 120$
$H_7(x) =$ $2^7 x^7 - 1344x^5 + 3360x^3 - 1680x$
$H_8(x) =$ $2^8 x^8 - 3584x^6 + 13440x^4 - 13440x^2 + 1680$
$H_9(x) =$ $2^9 x^9 - 9216x^7 + 48384x^5 - 80640x^3 + 30240x$
$H_{10}(x) =$ $2^{10} x^{10} - 23040x^8 + 161280x^6 - 403200x^4 + 302400x^2 - 30240$

Comparing the same power of t, we get Hermite polynomials,

$$H_0(x) = 1$$

$$H_1(x) = 2x$$

$$H_2(x) = 4x^2 - 2.$$

One can obtain an important relation of Hermite function by replacing the x and t by $-x$ and $-t$, and we get

$$g(-x, -t) = e^{-t^2 + 2tx} = \sum_{n=0}^{\infty} H_n(-x) \frac{(-t)^n}{n!}. \tag{5.14}$$

$$= (-1)^n \sum_{n=0}^{\infty} H_n(-x) \frac{t^n}{n!} \tag{5.15}$$

From (5.13) and (5.15), we get

$$H_n(x) = (-1)^n H_n(-x).$$

5.3 Recurrence Relations

Since $g(x, t)$ is a function of two variables x and t, if we partially differentiate $g(x, t)$ with respect to t and x, respectively, we get two

5.3 Recurrence Relations

recurrence relations. Differentiating (5.13) w.r.t. t we get

$$e^{-t^2+2tx}(-2t+2x) = \sum_{n=0}^{\infty} H_n(x)\, n\frac{t^{n-1}}{n!}. \qquad (5.16)$$

Using (5.13),

$$\sum_{n=0}^{\infty} H_n(x)\frac{t^n}{n!}(-2t+2x) = \sum_{n=0}^{\infty} H_n(x)\, n\frac{t^{n-1}}{n!} \qquad (5.17)$$

$$-2t\sum_{n=0}^{\infty} H_n(x)\frac{t^n}{n!} + 2x\sum_{n=0}^{\infty} H_n(x)\frac{t^n}{n!} = \sum_{n=0}^{\infty} H_n(x)\, n\frac{t^{n-1}}{n!} \qquad (5.18)$$

$$-2\sum_{n=0}^{\infty} H_n(x)\frac{t^{n+1}}{n!} + 2x\sum_{n=0}^{\infty} H_n(x)\frac{t^n}{n!} = \sum_{n=0}^{\infty} H_n(x)\, n\frac{t^{n-1}}{n!}. \qquad (5.19)$$

To make the equation in the same power of t, i.e., t^n, we have to replace n by $(n-1)$ and $(n+1)$ in left and right-hand side of (5.19), respectively. We get

$$-2\sum_{n=0}^{\infty} H_{n-1}(x)\frac{t^n}{(n-1)!} + 2x\sum_{n=0}^{\infty} H_n(x)\frac{t^n}{n!}$$
$$= \sum_{n=0}^{\infty} H_{n+1}(x)\,(n+1)\frac{t^n}{(n+1)!}. \qquad (5.20)$$

By using n! = n (n − 1)! *and* (n + 1)! = (n + 1) n!, (5.20) becomes

$$-2\sum_{n=0}^{\infty} nH_{n-1}(x)\frac{t^n}{n!} + 2x\sum_{n=0}^{\infty} H_n(x)\frac{t^n}{n!} = \sum_{n=0}^{\infty} H_{n+1}(x)\frac{t^n}{n!} \qquad (5.21)$$

$$\sum_{n=0}^{\infty} \frac{t^n}{n!}\left(-2nH_{n-1}(x) + 2xH_n(x) - H_{n+1}(x)\right) = 0. \qquad (5.22)$$

72 Hermite Polynomials

(5.22) tells us that each cofficient of t^n must vanishes. Then we get our first recurrence relation

$$H_{n+1}(x) = 2xH_n(x) - 2nH_{n-1}(x). \tag{5.23}$$

We can also find the Hermite polynomial using this recurrence relation.

Similarly, differentiating (5.13) w.r.t. x we get

$$e^{-t^2+2tx}(2t) = \sum_{n=0}^{\infty} H'_n(x)\frac{t^n}{n!} \tag{5.24}$$

Again using (5.13),

$$2t\sum_{n=0}^{\infty} H_n(x)\frac{t^n}{n!} = \sum_{n=0}^{\infty} H'_n(x)\frac{t^n}{n!} \tag{5.25}$$

$$2\sum_{n=0}^{\infty} H_n(x)\frac{t^{n+1}}{n!} = \sum_{n=0}^{\infty} H'_n(x)\frac{t^n}{n!} \tag{5.26}$$

$$2\sum_{n=0}^{\infty} H_{n-1}(x)\frac{t^n}{(n-1)!} = \sum_{n=0}^{\infty} H'_n(x)\frac{t^n}{n!} \tag{5.27}$$

$$2\sum_{n=0}^{\infty} nH_{n-1}(x)\frac{t^n}{n!} = \sum_{n=0}^{\infty} H'_n(x)\frac{t^n}{n!} \tag{5.28}$$

$$\sum_{n=0}^{\infty} \frac{t^n}{n!}\left(2nH_{n-1}(x) - H'_n(x)\right) = 0 \tag{5.29}$$

$$H'_n(x) = 2nH_{n-1}(x). \tag{5.30}$$

In terms of $H_n(x)$, replacing $n \to (n+1)$, we get

$$H'_{n+1}(x) = 2(n+1)H_n(x). \tag{5.31}$$

5.4 Rodrigues Formula

We can rewrite the generating function in the form

$$g(x,t) = e^{x^2} e^{-x^2} e^{-t^2+2tx} = \sum_{n=0}^{\infty} H_n(x) \frac{t^n}{n!} \tag{5.32}$$

$$g(x,t) = e^{x^2} e^{-(t-x)^2} \tag{5.33}$$

and we know that for $e^{-(t-x)^2}$

$$\frac{\partial}{\partial t} e^{-(t-x)^2} \implies -2(t-x)e^{-(t-x)^2} \tag{5.34}$$

$$\frac{\partial}{\partial x} e^{-(t-x)^2} \implies 2(t-x)e^{-(t-x)^2}. \tag{5.35}$$

From the above two expressions, we say that

$$\frac{\partial}{\partial t} e^{-(t-x)^2} = -\frac{\partial}{\partial x} e^{-(t-x)^2} \tag{5.36}$$

$$\frac{\partial^n}{\partial t^n} e^{-(t-x)^2} = (-1)^n \frac{\partial^n}{\partial x^n} e^{-(t-x)^2}. \tag{5.37}$$

Taking the partial n^{th} *derivative* of (5.13) and taking the limit $t \to 0$, we get

$$lim_{t \to 0} \frac{\partial^n}{\partial t^n} (e^{-t^2+2xt}) = lim_{t \to 0} \frac{\partial^n}{\partial t^n} \left(\sum_{n=0}^{\infty} H_n(x) \frac{t^n}{n!} \right)$$

$$lim_{t \to 0} \frac{\partial^n}{\partial t^n} (e^{-t^2+2xt}) = lim_{t \to 0} \frac{\partial^n}{\partial t^n}$$

$$\left(H_0(x) + H_1(x)\frac{t}{1!} + ... + H_n(x)\frac{t^n}{n!} + H_{n+1}(x)\frac{t^{n+1}}{(n+1)!} + ... \right)$$

74 Hermite Polynomials

$$lim_{t \to 0} \frac{\partial^n}{\partial t^n}(e^{-t^2+2xt}) = lim_{t \to 0} \frac{\partial^n}{\partial t^n}$$

$$\left(1 + H_1(x)\frac{t}{1!} + \ldots + H_n(x)\frac{t^n}{n!} + H_{n+1}(x)\frac{t^{n+1}}{(n+1)!} + \ldots\right)$$

$$lim_{t \to 0} \frac{\partial^n}{\partial t^n}(e^{-t^2+2xt}) = lim_{t \to 0}$$

$$\left(H_n(x) + \frac{(n+1)n(n-1)\ldots 2}{(n+1)!}H_{n+1}(x).t + \ldots\right)$$

$$lim_{t \to 0} \frac{\partial^n g}{\partial t^n} = H_n(x)$$

Now,
$$H_n(x) = lim_{t \to 0} \frac{\partial^n g}{\partial t^n},$$

Using (5.33),
$$H_n(x) = lim_{t \to 0} e^{x^2} \frac{\partial^n}{\partial t^n} e^{-(t-x)^2}.$$

Again using (5.37), we get
$$H_n(x) = lim_{t \to 0} e^{x^2} (-1)^n \frac{\partial^n}{\partial x^n} e^{-(t-x)^2}$$

$$\boldsymbol{H_n(x) = (-1)^n e^{x^2} \frac{\partial^n}{\partial x^n} e^{-x^2}}. \qquad (5.38)$$

The above expression is the Rodrigues formula.

5.5 Orthogonality and Normalilty

The orthogonality of the Hermite function is defined by a multiplicative weight factor $u = e^{-x^2}$, i.e.,

$$\int_{-\infty}^{+\infty} H_m(x) H_n(x) e^{-x^2} dx = 0 \quad if \ m \neq n \qquad (5.39)$$

5.5 Orthogonality and Normalilty

and the normality is defined as,

$$\int_{-\infty}^{+\infty} H_m(x)H_n(x)e^{-x^2}\,dx = 2^n n! \pi^{1/2} \quad if\ m = n.$$

(5.40)

We know the Rodrigues formula

$$H_n(x) = (-1)^n e^{x^2} \frac{d^n}{dx^n}(e^{-x^2}).$$

Since m and n is a non-negative integer, then the integration limits have changed and (5.40) becomes

$$\int_{0}^{+\infty} H_m(x)H_n(x)e^{-x^2}\,dx = 2^n n! \pi^{1/2} \quad if\ m = n.$$

(5.41)

If $m = n$, and using Rodrigues formula

$$\implies (-1)^n \int_{0}^{+\infty} \frac{d^n}{dx^n}(e^{-x^2}) H_n(x)\,dx.$$

(5.42)

Integrating (5.42) using integration by parts and taking $H_n(x)$ as the first function, we get

$$\implies \left(H_n(x)e^{-x^2}\big|_0^\infty + \int_0^\infty \frac{dH_n}{dx^n} e^{-x^2}\,dx \right)$$

(5.43)

The first term of the above expression goes to zero and using $\frac{dH_n}{dx^n} = 2^n n!$ property of Hermite function, we get

$$\implies 2^n n! \int_0^\infty e^{-x^2}\,dx.$$

We know the integration value, i.e., $\int_0^\infty e^{-x^2}\,dx = \pi^{1/2}$,

$$\implies 2^n n! \pi^{1/2}.$$

76 Hermite Polynomials

Equations (5.39) and (5.40) are together termed as orthonormality, and it is defined as

$$\int_{-\infty}^{+\infty} H_m(x) H_n(x) e^{-x^2}\, dx = \begin{cases} 0 & If,\ m \neq n \\ 2n!\pi^{1/2} & If,\ m = n \end{cases}$$

5.6 Application to the Simple Harmonic Oscillator

Consideration of the harmonic oscillator is very much important because of many continuous physical system governed by the harmonic oscillator. In microscopic level or quantum realm, atoms and electrons are not stationary; indeed, they are always in a mode of vibrations. Such types of vibrations can be considered as the superposition of an infinite number of simple harmonic oscillator. In quantum physics, to deal with properties of microscopic particles like wave functions and energy, we have a very famous second-order homogeneous differential equation known as Schrodinger equation. It is given by

$$\frac{\hbar^2}{2m} \nabla^2 \psi(x) + (E - V(x)) \psi(x) = 0. \tag{5.44}$$

The harmonic potential is given by

$$V(x) = \frac{1}{2} m \omega^2 x^2. \tag{5.45}$$

Substituting this potential in the Schrodinger (5.44), we get

$$\frac{\hbar^2}{2m} \nabla^2 \psi(x) + \left(E - \frac{1}{2} m \omega^2 x^2 \right) \psi(x) = 0 \tag{5.46}$$

$$\nabla^2 \psi(x) + \frac{2m}{\hbar^2} \left(E - \frac{1}{2} m \omega^2 x^2 \right) \psi(x) = 0. \tag{5.47}$$

5.6 Application to the Simple Harmonic Oscillator

We define a quantity χ,

$$\chi = \left(\frac{m\omega}{\hbar}\right)^{1/2} x \qquad (5.48)$$

$$\frac{d\chi}{dx} = \left(\frac{m\omega}{\hbar}\right)^{1/2}.$$

Now, we can rewrite $\nabla^2 \psi(x)$ in terms of χ

$$\frac{d\psi}{d\chi}\frac{d\chi}{dx} \Longrightarrow \left(\frac{m\omega}{\hbar}\right)^{1/2}\frac{d\psi}{d\chi}$$

$$\frac{d^2\psi}{d\chi^2} = \frac{d}{d\chi}\left(\frac{d\psi}{d\chi}\right)\frac{d\chi}{dx} \Longrightarrow \left(\frac{m\omega}{\hbar}\right)\frac{d^2\psi}{d\chi^2}$$

substituting these differentials in (5.47), we get

$$\left(\frac{m\omega}{\hbar}\right)\frac{d^2\psi}{d\chi^2} + \frac{2m}{\hbar^2}\left(E - \frac{1}{2}m\omega^2 x^2\right)\psi(\chi) = 0$$

$$\frac{d^2\psi}{d\chi^2} + \left(\frac{2E}{\hbar\omega} - \chi^2\right)\psi(\chi) = 0. \qquad (5.49)$$

Now consider a solution of the above equation for $\chi \to \infty$ and $\chi \to 0$. When $\chi \to \infty$, (5.49) becomes

$$\frac{d^2\psi}{d\chi^2} + \chi^2 \psi(\chi) = 0. \qquad (5.50)$$

The solution of above equation has the form

$$\psi = e^{\pm \chi^2/2} \qquad (5.51)$$

and for $\chi \to 0$, (5.49) becomes

$$\frac{d^2\psi}{d\chi^2} + \left(\frac{2E}{\hbar\omega}\right)\psi(\chi) = 0.$$

Again, the solution of the above equation has the form

$$\psi = A + B\chi + C\chi^2$$

78 Hermite Polynomials

$$\psi \sim B\chi. \tag{5.52}$$

From these two behaviors of ψ defined in (5.51) and (5.52), we can try a general solution that has the form

$$\psi = \varphi(\chi)e^{-\chi^2/2}. \tag{5.53}$$

If this is the solution we considered, then it must satisfy (5.49), i.e.,

$$\psi'' = \varphi'' e^{-\chi^2/2} - 2\chi\varphi' e^{-\chi^2/2} - \chi\varphi' e^{-\chi^2/2} - \chi^2\varphi e^{-\chi^2/2}. \tag{5.54}$$

From (5.50), (5.53), and (5.54), we get

$$\varphi'' e^{-\chi^2/2} - 2\chi\varphi' e^{-\chi^2/2} - \chi\varphi' e^{-\chi^2/2} - \chi^2\varphi e^{-\chi^2/2} + \chi^2\varphi e^{-\chi^2/2} = 0$$

$$\frac{d^2\varphi}{d\chi^2} - 2\chi\frac{d\varphi}{d\chi} + 2\chi\varphi = 0. \tag{5.55}$$

The above equation is the same as (5.1); so the solutions of above the equation are the Hermite polynomials, i.e.,

$$\psi_n = H_n(\chi). \tag{5.56}$$

and the general solution of (5.47) by using (5.53) and (5.56) is written as

$$\psi = H_n(x)e^{-\chi^2/2}$$

where $H_n(x)$ is our Hermite polynomials.

5.7 Exercises

1. Prove the orthogonality of Hermite function defined in Section 5.5.

2. Given a function $\psi(x) = NH_n(x)e^{-x^2/2}$, where N is a normalization constant. Check that the value of N is $\left(1/2^n n! \pi^{1/2}\right)^{1/2}$. Hint: use the expression $\int_{-\infty}^{+\infty} \psi_m(x) \psi_n(x) \mathrm{d}x = 1$.

3. Using recurrence relation of (5.23) and (5.30), show that $H_n''(x) - 2xH_n'(x) + 2nH_n(x) = 0$.

4. Prove the following relations:

   a) $H_n(x) = 2xH_{n-1}(x) - 2(n-1)H_{n-2}(x) = 0$
   b) $\quad xH_n'(x) = nH_{n-1}'(x) + nH_n(x) = 0$.

5. Prove that

   a) $H_{2n+1}(0) = 0$
   b) $H_{2n}'(0) = 0$.

6

Laguerre Polynomials

6.1 Laguerre Functions

Laguerre function is the solution of second-order differential equation that has the form

$$x\frac{d^2y}{dx^2} + (1-x)\frac{dy}{dx} + ny = 0 \tag{6.1}$$

and the equation has an essential singularity at $x = \infty$. $and\, n$ is any integer. Since $x = 0$ is a regular singularity of the equation. We can find the complementary function of the above equation by using a most generalized power series or Frobenius series:

$$y = \sum_{k=0}^{\infty} a_k x^{k+p}. \tag{6.2}$$

Taking the first and second derivative of (6.2), we get

$$\frac{dy}{dx} = \sum_{k=0}^{\infty} a_k(k+p)x^{k+p-1} \tag{6.3}$$

$$\frac{d^2y}{dx^2} = \sum_{k=0}^{\infty} a_k(k+p)(k+p-1)x^{k+p-2}. \tag{6.4}$$

Putting (6.2), (6.3), and (6.4) into (6.1), we get

$$\sum_{k=0}^{\infty} a_k(k+p)(k+p-1)x^{k+p-1} + (1-x)\sum_{k=0}^{\infty} a_k(k+p)x^{k+p-1} + n\sum_{k=0}^{\infty} a_k x^{k+p} = 0. \tag{6.5}$$

82   Laguerre Polynomials

Multiplying the above by x^{-p+1}, we get

$$\sum_{k=0}^{\infty} a_k(k+p)(k+p-1)x^k + (1-x)\sum_{k=0}^{\infty} a_k(k+p)x^k + nx\sum_{k=0}^{\infty} a_k x^k = 0 \tag{6.6}$$

$$\sum_{k=0}^{\infty} a_k x^k \left((k+p)(k+p-1) + (1-x)(k+p) + nx\right) = 0 \tag{6.7}$$

$$\sum_{k=0}^{\infty} a_k x^k \left((k+p)(k+p-1) + (k+p) - (k+p)x + nx\right) = 0 \tag{6.8}$$

$$\sum_{k=0}^{\infty} a_k x^k \left((k+p)(k+p)\right) - (k+p-n)\sum_{k=0}^{\infty} a_k x^{k+1} = 0 \tag{6.9}$$

Taking x^0 coefficient of (6.9) we obtain,

$$p^2 = 0. \tag{6.10}$$

Equation (6.10) is called the indicial equation and it has two repeated roots, i.e., $p = 0$. Now (6.9) gives

$$\sum_{k=0}^{\infty} \left((k+p)(k+p)a_k x^k - (k+p-n)a_k x^{k+1}\right). \tag{6.11}$$

Making (6.11) in x^k by changing $k \to k-1$ in the second term, we get

$$\sum_{k=0}^{\infty} \left((k+p)(k+p)a_k x^k - (k-1+p-n)a_{k-1} x^k\right) \tag{6.12}$$

$$\sum_{k=0}^{\infty} x^k \left((k+p)(k+p)a_k - (k-1+p-n)a_{k-1}\right). \tag{6.13}$$

Again, (6.13) demands that each cofficients of x^k must vanish. We get

$$a_k(k+p)(k+p) - (k-1+p-n)a_{k-1} = 0. \tag{6.14}$$

6.1 Laguerre Functions

Taking root of (6.10), i.e., $p = 0$, we get our recurrence relation

$$a_k = \frac{(k-1-n)}{k^2} a_{k-1}, \qquad (6.15)$$

$k = 1, 2, 3, 4, 5, ...$

Now, take the integer values of k

$$k = 1, \qquad a_1 = -na_0,$$

$$k = 2 \qquad a_2 = \frac{(n-1)}{2^2}.na_0,$$

$$k = 3, \qquad a_3 = -\frac{(n-2)}{3^2}.\frac{(n-1)}{2^2}.na_0,$$

$$k = 4, \qquad a_4 = \frac{(n-3)}{4^2}.\frac{(n-2)}{3^2}.\frac{(n-1)}{2^2}.na_0,$$

$$k = 5, \qquad a_5 = -\frac{(n-4)}{5^2}.\frac{(n-3)}{4^2}.\frac{(n-2)}{3^2}.\frac{(n-1)}{2^2}.na_0,$$

and so on. The explicit form of (6.2) can be written as

$$y = a_0 + a_1 x + a_2 x^2 + a_3 x^3 + a_4 x^4 + a_5 x^5 + a_6 x^6 + a_7 x^7 + ... \qquad (6.16)$$

Putting the value of $a_0, a_1, a_2, a_3, ...$ in (6.16), we get

$$y = a_0 \left(1 - nx + \frac{(n-1)}{2^2}.n.x^2 - \frac{(n-2)}{3^2}.\frac{(n-1)}{2^2}.n.x^3 \right.$$

$$\left. + \frac{(n-3)}{4^2}.\frac{(n-2)}{3^2}.\frac{(n-1)}{2^2}.n.x^4 - ... \right) \qquad (6.17)$$

(6.17) is our solution of second-order differential (6.1). Now, taking the integer values of n, our solution y becomes a polynomial termed as Laguerre polynomials.

$$n = 0, \quad y = a_0,$$

$$n = 1, \quad y = a_0(1-x),$$

$$n = 2 \quad y = a_0(1 - 2x + \frac{x^2}{2}),$$

$$n = 3, \quad y = a_0(1 - 3x + \frac{3x^2}{2} - \frac{3x^3}{18})$$

$$n = 4, \quad y = a_0(1 - 4x + 3x^2 - \frac{2x^3}{3} + \frac{x^4}{24}),$$

$$n = 5, \quad y = a_0(1 - 5x + 5x^2 - \frac{5x^3}{3} + \frac{5x^4}{24} - \frac{5x^5}{600}).$$

Now, choosing the value of $a_0 = 1$, we get Laguerre polynomials

$$L_0(x) = 1,$$

$$L_1(x) = (1-x),$$

$$L_2(x) = \frac{1}{2!}(x^2 - 4x + 2),$$

$$L_3(x) = \frac{1}{3!}(-x^3 + 9x^2 - 18x + 6),$$

$$L_4(x) = \frac{1}{4!}(x^4 - 16x^3 + 72x^2 - 96x + 24),$$

$$L_5(x) = \frac{1}{5!}(-x^5 + 25x^4 - 200x^3 + 600x^2 - 600x + 120)$$

and so on, and you can find the Laguerre polynomials. Since Laguerre polynomials $L_n(x)$ are solutions of (6.1), Hence, it satisfies(6.1), and we get

$$xL_n''(x) + (1-x)L_n'(x) + nL_n(x) = 0. \tag{6.18}$$

The general expression or n^{th} term of Laguerre polynomial is given by,

$$L_n(x) = \sum_{m=0}^{n} (-1)^m \frac{n!}{(m!)^2 (n-m)!} (x)^m.$$

We can generates the Laguerre polynomial using the above expression by terminating a series for integral n.

6.2 Generating Function

The generating function of Laguerre function $L_n(x)$ is defined as

$$g(x,t) = \frac{e^{-xt/(1-t)}}{1-t} = \sum_{n=0}^{\infty} L_n(x) t^n. \tag{6.19}$$

We know that

$$e^x = \sum_{n=0}^{\infty} \frac{x^n}{n!} = 1 + \frac{x}{1!} + \frac{x^2}{2!} + \frac{x^3}{3!} + \ldots$$

Now, the left-hand side of (6.19) has the form

$$\frac{e^{-xt/(1-t)}}{1-t} = \sum_{n=0}^{\infty} \frac{(-xt/(1-t))^n}{n!(1-t)}$$

$$\sum_{n=0}^{\infty} \frac{(-xt/(1-t))^n}{n!(1-t)} = \frac{1}{(1-t)}$$
$$\times \left(1 + \frac{(-xt/(1-t))}{1!} + \frac{(-xt/(1-t))^2}{2!} + \frac{(-xt/(1-t))^3}{3!} + \ldots\right)$$

86 Laguerre Polynomials

Putting the above series in (6.19), we get

$$\frac{1}{(1-t)}\left(1 + \frac{(-xt/(1-t))}{1!} + \frac{(-xt/(1-t))^2}{2!} + \frac{(-xt/(1-t))^3}{3!} + \ldots\right)$$

$$= \sum_{n=0}^{\infty} L_n(x) t^n$$

$$\frac{1}{(1-t)}\left(1 + \frac{(-xt)}{1!(1-t)} + \frac{(-xt)^2}{2!(1-t)^2} + \ldots\right) = L_0(x) + L_1(x)t + L_2(x)t^2 + \ldots$$

Multipling the above equation by $(1-t)$, we get

$$1 - \frac{xt}{1!(1-t)} + \frac{x^2 t^2}{2!(1-t)^2} + \ldots = (1-t)\left(L_0(x) + L_1(x)t + L_2(x)t^2 + \ldots\right)$$

$$1 - \frac{xt}{1!(1-t)} + \frac{x^2 t^2}{2!(1-t)^2} + \ldots = L_0(x) - L_0(x)t$$

$$+ L_1(x)t - L_1(x)t^2 + L_2(x)t^2 - L_2(x)t^3 + \ldots$$

$$1 - \frac{xt(1-t)^{-1}}{1!} + \frac{x^2 t^2 (1-t)^{-2}}{2!} + \ldots = L_0(x) - L_0(x)t$$

$$+ L_1(x)t - L_1(x)t^2 + L_2(x)t^2 - L_2(x)t^3 + \ldots$$

Using the binomial series, $(1-t)^{-n} = 1 + nt + n(n-1)t^2 + \ldots$

$$1 - \frac{xt(1+t)}{1!} + \frac{x^2 t^2 (1+2t)}{2!} + \ldots$$

$$= L_0(x) - L_0(x)t + L_1(x)t - L_1(x)t^2 + L_2(x)t^2 - L_2(x)t^3 + \ldots$$

$$1 - \frac{(xt + xt^2)}{1!} + \frac{(x^2 t^2 + 2x^2 t^3)}{2!} + \ldots$$

$$= L_0(x) + (L_1(x) - L_0(x))\, t + (L_2(x) - L_1(x))\, t^2 +$$

$$+ (L_2(x) - L_2(x))\, t^3 + \dots$$

$$1 - xt + t^2 \left(-\frac{x}{1!} + \frac{x^2}{2!}\right) + \dots = L_0(x) + (L_1(x)$$

$$- L_0(x))t + (L_2(x) - L_1(x))\, t^2 + (L_2(x) - L_2(x))\, t^3 + \dots$$

Comparing the same power of t, we get Laguerre polynomials

$$L_0(x) = 1$$

$$L_1(x) - L_0(x) = -x,$$

$$\implies L_1(x) = 1 - x$$

$$L_2(x) - L_1(x) = -\frac{x}{1!} + \frac{x^2}{2!} \implies L_2(x) = -\frac{x}{1!} + \frac{x^2}{2!} + L_1(x),$$

$$\implies L_2(x) = 1 - x - \frac{x}{1!} + \frac{x^2}{2!}$$

$$\implies L_2(x) = \frac{1}{2!}(x^2 - 4x + 2).$$

6.3 Recurrence Relations

Since $g(x,t)$ is a function of two variables x and t, if we partially differentiate $g(x,t)$ with respect to t and x, respectively, we get two recurrence relations.

$$\frac{e^{-xt/(1-t)}}{(1-t)} = \sum_{n=0}^{\infty} L_n(x) t^n. \qquad (6.20)$$

Table 6.1 Some Laguerre Polynomials

$L_0(x) = 1$

$L_1(x) = (1 - x)$

$L_2(x) = \frac{1}{2!}(x^2 - 4x + 2)$

$L_3(x) = \frac{1}{3!}(-x^3 + 9x^2 - 18x + 6)$

$L_4(x) = \frac{1}{4!}(x^4 - 16x^3 + 72x^2 - 96x + 24)$

$L_5(x) = \frac{1}{5!}(-x^5 + 25x^4 - 200x^3 + 600x^2 - 600x + 120)$

$L_6(x) = \frac{1}{6!}(x^6 - 36x^5 + 450x^4 - 2400x^3 + 5400x^2 - 4320x + 720)$

$L_7(x) = \frac{1}{7!}(-x^7 + 49x^6 - 882x^5 + 7350x^4 - 29400x^3 + 52920x^2 - 35280x + 5040)$

$L_8(x) = \frac{1}{8!}(x^8 - 64x^7 + 1568x^6 - 18816x^5 + 117600x^4 - 376320x^3 + 564480x^2 - 322560x + 40320)$

6.3 Recurrence Relations

Differentiating (6.20) w.r.t. t, we get,

$$e^{-xt/(1-t)}\frac{1}{(1-t)^2} + \frac{(-x)e^{-xt/(1-t)}}{(1-t)}\frac{d}{dx}\frac{t}{(1-t)} = n\sum_{n=0}^{\infty} L_n(x)t^{n-1} \quad (6.21)$$

$$e^{-xt/(1-t)}\frac{1}{(1-t)^2} - \frac{xe^{-xt/(1-t)}}{(1-t)}\frac{1}{(1-t)^2} = n\sum_{n=0}^{\infty} L_n(x)t^{n-1}. \quad (6.22)$$

Using (6.20),

$$\sum_{n=0}^{\infty} L_n(x)t^n \frac{1}{(1-t)} - \sum_{n=0}^{\infty} L_n(x)t^n \frac{x}{(1-t)^2} = n\sum_{n=0}^{\infty} L_n(x)t^{n-1} \quad (6.23)$$

$$\sum_{n=0}^{\infty} L_n(x)t^n(1-t) - x\sum_{n=0}^{\infty} L_n(x)t^n = n\sum_{n=0}^{\infty} L_n(x)t^{n-1}(1-t)^2 \quad (6.24)$$

$$\sum_{n=0}^{\infty} L_n(x)t^n(1-t) - x\sum_{n=0}^{\infty} L_n(x)t^n = n\sum_{n=0}^{\infty} L_n(x)t^{n-1}(1+t^2-2t) \quad (6.25)$$

$$\sum_{n=0}^{\infty} L_n(x)t^n - \sum_{n=0}^{\infty} L_n(x)t^{n+1} - x\sum_{n=0}^{\infty} L_n(x)t^n$$

$$= n\sum_{n=0}^{\infty} L_n(x)t^{n-1} + n\sum_{n=0}^{\infty} L_n(x)t^{n+1} - 2n\sum_{n=0}^{\infty} L_n(x)t^n \quad (6.26)$$

$$-(n+1)\sum_{n=0}^{\infty} L_n(x)t^{n+1} = n\sum_{n=0}^{\infty} L_n(x)t^{n-1} - (2n-x+1)\sum_{n=0}^{\infty} L_n(x)t^n. \quad (6.27)$$

To make the equation in the same power of t, i.e., t^n, we have to replace n by $(n-1)$ and $(n+1)$ in left and right-hand side of (6.27), respectively. We get

$$-n\sum_{n=0}^{\infty} L_{n-1}(x)t^n = (n+1)\sum_{n=0}^{\infty} L_{n+1}(x)t^n - (2n-x+1)\sum_{n=0}^{\infty} L_n(x)t^n \quad (6.28)$$

90 Laguerre Polynomials

$$(n+1)\sum_{n=0}^{\infty} L_{n+1}(x)t^n - (2n-x+1)\sum_{n=0}^{\infty} L_n(x)t^n + n\sum_{n=0}^{\infty} L_{n-1}(x)t^n = 0 \quad (6.29)$$

$$\sum_{n=0}^{\infty} t^n \left((n+1)L_{n+1}(x) - (2n-x+1)L_n(x) + nL_{n-1}(x)\right) = 0. \quad (6.30)$$

(6.30) tells us that each coefficient of t^n must vanishes. Then we get our first recurrence relation

$$(n+1)L_{n+1}(x) = (2n-x+1)L_n(x) - nL_{n-1}(x). \quad (6.31)$$

We can also find the Laguerre polynomial using this recurrence relation.

Similarly, differentiating (6.20) w.r.t. x we get

$$e^{-xt/(1-t)} \frac{-t}{(1-t)^2} = \sum_{n=0}^{\infty} L'_n(x)t^n. \quad (6.32)$$

Again using (6.20),

$$-\frac{t}{(1-t)} \sum_{n=0}^{\infty} L_n(x)t^n = \sum_{n=0}^{\infty} L'_n(x)t^n \quad (6.33)$$

$$-\sum_{n=0}^{\infty} L_n(x)t^{n+1} = \sum_{n=0}^{\infty} L'_n(x)t^n(1-t) \quad (6.34)$$

$$-\sum_{n=0}^{\infty} L_n(x)t^{n+1} = \sum_{n=0}^{\infty} L'_n(x)t^n - \sum_{n=0}^{\infty} L'_n(x)t^{n+1} \quad (6.35)$$

$$-\sum_{n=0}^{\infty} L_{n-1}(x)t^n = \sum_{n=0}^{\infty} L'_n(x)t^n - \sum_{n=0}^{\infty} L'_{n-1}(x)t^n \quad (6.36)$$

$$\sum_{n=0}^{\infty} t^n \left(L'_n(x) + L_{n-1}(x) - L'_{n-1}(x)\right) = 0 \quad (6.37)$$

$$L'_{n-1}(x) = L'_n(x) + L_{n-1}(x). \qquad (6.38)$$

Replace n by $n+1$ in (6.38)

$$L'_{n+1}(x) = L'_n(x) - L_n(x). \qquad (6.39)$$

Now, differentiating (6.31) w.r.t. to x, we get

$$(n+1)L'_{n+1}(x) = (2n - x + 1)L'_n(x) - L_n(x) - nL'_{n-1}(x). \qquad (6.40)$$

Using (6.38), (6.39), and (6.40), we get

$$(n+1)\left(L'_n(x) - L_n(x)\right) = (2n - x + 1)L'_n(x) - L_n(x) - n\left(L'_n(x) + L_{n-1}(x)\right) \qquad (6.41)$$

$$(n+1)L'_n(x) - (n+1)L_n(x) = 2nL'_n(x) - xL'_n(x)$$
$$+ L'_n(x) - L_n(x) - nL'_n(x) - nL_{n-1}(x).$$

After rearranging terms, we get

$$xL'_n(x) = nL_n(x) - nL_{n-1}(x). \qquad (6.42)$$

(6.31), (6.38), and (6.42) are our desired recurrence relations.

6.4 Rodrigues Formula

Rodrigues formula for Laguerre polynomial is defined as

$$L_n(x) = \frac{e^x}{n!} \frac{d^n}{dx^n}(x^n e^{-x}).$$

To prove the above expression, we have to do the nth derivative of two variables. For this, we have used the Leibniz theorem that is given by

$$\frac{d^n}{dx^n}(uv) = \sum_{q=0}^{n} {}^nC_q \frac{d^{n-q}u}{dx^{n-q}} \frac{d^q v}{dx^q}.$$

Hence,

$$\Rightarrow \frac{e^x}{n!} \sum_{q=0}^{n} {}^nC_q \frac{d^q x^n}{dx^q} \frac{d^{n-q} e^{-x}}{dx^{n-q}}$$

$$\Rightarrow \frac{e^x}{n!} \sum_{q=0}^{n} \frac{n!}{q!(n-q)!} \frac{n!}{(n-q)!} x^{n-q} \frac{d^{n-q} e^{-x}}{dx^{n-q}}$$

$$\Rightarrow \frac{e^x}{n!} \sum_{q=0}^{n} \frac{n!}{q!(n-q)!} \frac{n!}{(n-q)!} x^{n-q} (-1)^{n-q} e^{-x}$$

$$\Rightarrow e^x \sum_{q=0}^{n} (-1)^{n-q} \frac{n!}{q!(n-q)!} \frac{1}{(n-q)!} x^{n-q} e^{-x}.$$

Changing the variable $(n-q) \to m$, we get

$$\Rightarrow \sum_{m=0}^{n} (-1)^m \frac{n!}{(m!)^2 (n-m)!} x^m$$

$$\Rightarrow L_m(x).$$

6.5 Orthonormality

The orthogonality of Laguerre function is defined by a multiplicative weight factor $u = e^{-x}$, i.e.,

$$\int_0^{+\infty} L_m(x) L_n(x) e^{-x} dx = 0 \quad if \ m \neq n \tag{6.43}$$

and the Normality is define as,

$$\int_0^{+\infty} L_m(x) L_n(x) e^{-x} dx = 1 \quad if \ m = n. \tag{6.44}$$

We know the generating function

$$\frac{e^{-xt/(1-t)}}{(1-t)} = \sum_{n=0}^{\infty} L_n(x) t^n \tag{6.45}$$

6.5 Orthonormality

and

$$\frac{e^{-xq/(1-q)}}{(1-q)} = \sum_{m=0}^{\infty} L_m(x)q^m. \tag{6.46}$$

Multiplying (6.45) and (6.46), we get

$$\sum_{m=0}^{\infty}\sum_{n=0}^{\infty} L_m(x)L_n(x)t^n q^m = \frac{e^{-xt/(1-t)}}{(1-t)}\frac{e^{-xq/(1-q)}}{(1-q)}. \tag{6.47}$$

Multiplying above equation by e^{-x},

$$\sum_{m=0}^{\infty}\sum_{n=0}^{\infty} \left(L_m(x)L_n(x)e^{-x}\right)t^n q^m = \frac{e^{-xt/(1-t)}}{(1-t)}\frac{e^{-xq/(1-q)}}{(1-q)}e^{-x}. \tag{6.48}$$

Integrating (6.48) w.r.t. x, we get

$$\int_0^{+\infty}\sum_{m=0}^{\infty}\sum_{n=0}^{\infty} \left(L_m(x)L_n(x)e^{-x}\right)t^n q^m dx = \int_0^{+\infty}\frac{e^{-xt/(1-t)}}{(1-t)}\frac{e^{-xq/(1-q)}}{(1-q)}e^{-x}dx \tag{6.49}$$

$$\sum_{m=0}^{\infty}\sum_{n=0}^{\infty}\left(\int_0^{+\infty} L_m(x)L_n(x)e^{-x}dx\right)t^n q^m = \int_0^{+\infty}\frac{e^{-x(t/(1-t)+q/(1-q)+1)}}{(1-t)(1-q)}dx \tag{6.50}$$

$$\implies -\frac{1}{(1-t)(1-q)}\frac{e^{-x(t/(1-t)+q/(1-q)+1)}}{\left(1+\frac{t}{(1-q)}+\frac{q}{(1-q)}\right)}\bigg|_0^{\infty}$$

$$\implies \frac{1}{(1-t)(1-q)}\frac{1}{\left(1+\frac{t}{(1-t)}+\frac{q}{(1-q)}\right)}$$

$$\implies \frac{1}{(1-qt)}$$

$$\implies (1-qt)^{-1} = 1 + qt + (qt)^2 + \ldots\ldots\ldots \tag{6.51}$$

$$\implies \sum_{n=0}^{\infty} q^n t^n. \tag{6.52}$$

Using (6.50) and (6.52), we get

$$\sum_{m=0}^{\infty}\sum_{n=0}^{\infty}\left(\int_{0}^{+\infty} L_m(x)L_n(x)e^{-x}dx\right)t^n q^m = \sum_{n=0}^{\infty} t^n q^n. \qquad (6.53)$$

If $m \neq n$, compare the coefficient of $t^n q^m$, and we get

$$\int_{0}^{+\infty} L_m(x)L_n(x)e^{-x}dx = 0 \qquad (6.54)$$

and for $m = n$,

$$\int_{0}^{+\infty} L_m(x)L_n(x)e^{-x}dx = 1. \qquad (6.55)$$

Equations (6.54) and eq.(6.55) are together termed as orthonormality, and it is defined as

$$\int_{0}^{+\infty} L_m(x)L_n(x)e^{-x}dx = \left\{ \begin{array}{ll} 0 & If,\ m \neq n \\ 1 & If,\ m = n \end{array} \right\}.$$

6.6 Application to the Hydrogen Atom

The important application of Laguerre function is the solution of hydrogen atom in quantum mechanics. Laguerre function comes in the solution of radial part of the wavefunction of hydrogen atom. Since the potential of hydrogen atom is spherically symmetric potential; hence, we use the spherical polar coordinate system (r, θ, ϕ). So the wavefunction is represented by

$$\psi(x) \to \psi(r, \theta, \phi),$$

$$\psi(r, \theta, \phi) = R(r)Y(\theta, \phi). \qquad (6.56)$$

The Schrodinger equation in spherical polar coordinate form is given by

$$\nabla^2 \psi(r, \theta, \phi) + \frac{2m}{\hbar^2}(E - V(r))\psi(r, \theta, \phi) = 0. \qquad (6.57)$$

6.6 Application to the Hydrogen Atom

Now, in spherical polar coordinates,

$$\nabla^2 \psi = \frac{1}{r^2}\frac{\partial \psi}{\partial r}\left(r^2 \frac{\partial \psi}{\partial r}\right) + \frac{1}{r^2}\left[\frac{1}{\sin\theta}\frac{\partial}{\partial \theta}\left(\sin\theta \frac{\partial \psi}{\partial \theta}\right) + \frac{1}{\sin^2\theta}\left(\frac{\partial^2 \psi}{\partial \phi^2}\right)\right]. \quad (6.58)$$

We also know at angular momentum operator L^2 with eigenvalue $l(l+1)$ has the form

$$L^2 = -\hbar^2\left[\frac{1}{\sin\theta}\frac{\partial}{\partial \theta}\left(\sin\theta \frac{\partial}{\partial \theta}\right) + \frac{1}{\sin^2\theta}\left(\frac{\partial^2}{\partial \phi^2}\right)\right]. \quad (6.59)$$

From (6.57), (6.58), and (6.59), we get

$$\frac{1}{r^2}\frac{\partial}{\partial r}\left(r^2 \frac{\partial \psi}{\partial r}\right) + \frac{2m}{\hbar^2}(E - V(r))\psi(r,\theta,\phi) = \frac{L^2 \psi}{\hbar^2 r^2}. \quad (6.60)$$

Putting (6.56) into (6.60) and then multiplying by $\frac{r^2}{R(r)Y(\theta,\phi)}$, we get

$$\frac{1}{R(r)}\frac{\partial}{\partial r}\left(r^2 \frac{\partial R}{\partial r}\right) + \frac{2mr^2}{\hbar^2}(E - V(r))R(r) = \frac{L^2 Y}{\hbar^2 Y(\theta,\phi)}. \quad (6.61)$$

(6.61) contains separately radial parts and angular part; so they must be equal to constant m:

$$\frac{1}{R(r)}\frac{\partial}{\partial r}\left(r^2 \frac{\partial R}{\partial r}\right) + \frac{2mr^2}{\hbar^2}(E - V(r))R(r) = \frac{L^2 Y}{\hbar^2 Y(\theta,\phi)} = m. \quad (6.62)$$

Now, we have

$$L^2 Y(\theta,\phi) = m\hbar^2 Y(\theta,\phi). \quad (6.63)$$

We know the eigenvalue of L^2 from the knowledge of quantum mechanics,

$$L^2 = l(l+1)\hbar. \quad (6.64)$$

Using (6.64) into (6.61), we get the radial part of schrodinger equation,

Laguerre Polynomials

$$\frac{d^2R}{dr^2} + \frac{2}{r}\frac{dR}{dr} + \frac{2m}{\hbar^2}\left(E - V(r) - \frac{l(l+1)\hbar^2}{2mr^2}\right)R(r) = 0. \quad (6.65)$$

The potential for hydrogen atom is coulomb potential and it is given by

$$V(r) = -\frac{e^2}{r}. \quad (6.66)$$

Substituting this potential in (6.65), we get

$$\frac{d^2R}{dr^2} + \frac{2}{r}\frac{dR}{dr} + \frac{2m}{\hbar^2}\left(E + \frac{e^2}{r} - \frac{l(l+1)\hbar^2}{2mr^2}\right)R(r) = 0. \quad (6.67)$$

We define a variable,

$$w(r) = rR(r).$$

Now (6.67) has the form

$$\frac{d^2w}{dr^2} + \frac{2m}{\hbar^2}\left(E + \frac{e^2}{r} - \frac{l(l+1)\hbar^2}{2mr^2}\right)w(r) = 0. \quad (6.68)$$

Now consider a solution of the above equation for $r \to \infty$ and $r \to 0$.
When $r \to \infty$, (6.68) becomes

$$\frac{d^2w}{dr^2} + \frac{2mE}{\hbar^2}w(r) = 0. \quad (6.69)$$

The solution of the above equation has the form

$$w = e^{\pm u}, \quad (where\ u = (-2mE/\hbar^2)^{1/2}). \quad (6.70)$$

Discarding the diverge solution,

$$w = e^{-u} \quad (6.71)$$

and for $r \to 0$, eq.(6.68) becomes

$$\frac{d^2w}{dr^2} - \left(\frac{l(l+1)\hbar^2}{2mr^2}\right)w(r) = 0.$$

6.6 Application to the Hydrogen Atom

Again, the solution of the above equation has the form

$$w = Ar^{l+1} + Br^{-l}.$$

Again discarding the diverge solution, i.e., at $r \to 0$, r^{-l} diverge then,

$$w \sim Ar^{l+1}. \tag{6.72}$$

From these two behavior of w defined in (6.71) and (6.72), we can try a general solution that has the form,

$$w = \varphi(r) r^{l+1} e^{-u}. \tag{6.73}$$

If this is the solution we considered, then it must satisfy (6.68), and we get

$$\frac{d^2 \varphi}{dr^2} + 2\left(\frac{l+1}{r} - u\right)\frac{d\varphi}{dr} + 2\left(\frac{-u(l+1) + me^2\hbar^2}{r}\right)\varphi(r) = 0. \tag{6.74}$$

The above equation is the same as (6.1); so the solutions of (6.74) are the Laguerre polynomials, i.e.,

$$R_{nl}(r) = N_{nl}\left(\frac{2r}{na_0}\right)^l e^{-r/na_0} L_{n+l}^{2l+1}\left(\frac{2r}{na_0}\right) \tag{6.75}$$

where N_{nl} is the normalization constant and is given by

$$N_{nl} = -\left(\frac{2}{na_0}\right)^{3/2}\left(\frac{(n-l-1)!}{2n\left((n+3)!\right)^3}\right)^{1/2} \tag{6.76}$$

where $L_{n+l}^{2l+1}(x)$ is an associated Laguerre polynomials. The associated Laguerre polynomial is the solution of equation that has the form

$$x\frac{d^2 y}{dx^2} + (m+1-x)\frac{dy}{dx} + ny = 0. \tag{6.77}$$

98 *Laguerre Polynomials*

$L_n^m(x)$ is the associated Laguerre polynomial which is the solution of (6.77) and it is related with Laguerre polynomial by the relation

$$L_n^m(x) = (-1)^m \frac{d^m}{dx^m} L_{n+m}(x) \tag{6.78}$$

where, $L_{n+m}(x)$ is our Laguerre polynomial.

6.7 Associated Laguerre Polynomials

Associated Laguerre function is the solution of generalized second-order differential equation that has the form

$$x\frac{d^2y}{dx^2} + (m+1-x)\frac{dy}{dx} + ny = 0. \tag{6.79}$$

For $m = 0$, we get the same (6.1)

Again, (6.79) has an essential singularity at $x = \infty$ and n and m is an any integer. Since $x = 0$ is a regular singularity of the equation. Here, we can also find the complementary function of the above equation by using the most generalized power series or Frobenius series:

$$y = \sum_{k=0}^{\infty} a_k x^{k+p}. \tag{6.80}$$

Taking the first and second derivative of (6.79), we get

$$\frac{dy}{dx} = \sum_{k=0}^{\infty} a_k (k+p) x^{k+p-1} \tag{6.81}$$

$$\frac{d^2y}{dx^2} = \sum_{k=0}^{\infty} a_k (k+p)(k+p-1) x^{k+p-2}. \tag{6.82}$$

6.7 Associated Laguerre Polynomials

Putting (6.80), (6.81), and (6.82) into (6.79), we get

$$\sum_{k=0}^{\infty} a_k(k+p)(k+p-1)x^{k+p-1} + (m+1-x)$$

$$\times \sum_{k=0}^{\infty} a_k(k+p)x^{k+p-1} + n \sum_{k=0}^{\infty} a_k x^{k+p} = 0 \quad (6.83)$$

Multiplying the above by x^{-p+1}, we get

$$\sum_{k=0}^{\infty} a_k(k+p)(k+p-1)x^k + (m+1-x)\sum_{k=0}^{\infty} a_k(k+p)x^k$$

$$+ nx \sum_{k=0}^{\infty} a_k x^k = 0 \quad (6.84)$$

$$\sum_{k=0}^{\infty} a_k x^k \left((k+p)(k+p-1) + (m+1-x)(k+p) + nx\right) = 0 \quad (6.85)$$

$$\sum_{k=0}^{\infty} a_k x^k \left((k+p)(k+p-1) + (m+1)(k+p) - (k+p)x + nx\right) = 0 \quad (6.86)$$

$$\sum_{k=0}^{\infty} a_k x^k \left((k+p)(k+p+m)\right) - (k+p-n)\sum_{k=0}^{\infty} a_k x^{k+1} = 0. \quad (6.87)$$

Taking x^0 coefficient of (6.87) we obtain,

$$p(p+m) = 0 \quad (6.88)$$

We have two roots, i.e., $p = 0$ and $p = -m$, Now (6.87) gives

$$\sum_{k=0}^{\infty} \left((k+p)(k+p+m)a_k x^k - (k+p-n)a_k x^{k+1}\right). \quad (6.89)$$

Making (6.89) in x^k by changing k $\to$ k $-$ 1 in the second term, we get

$$\sum_{k=0}^{\infty} \left((k+p)(k+p+m)a_k x^k - (k-1+p-n)a_{k-1}x^k\right) \quad (6.90)$$

100 Laguerre Polynomials

$$\sum_{k=0}^{\infty} x^k \left((k+p)(k+p+m)a_k - (k-1+p-n)a_{k-1}\right). \quad (6.91)$$

Again, (6.91) demands that each cofficients of x^k must vanish. We get,.

$$a_k(k+p)(k+p+m) - (k-1+p-n)a_{k-1} = 0. \quad (6.92)$$

Taking the non-negative root of (6.88), i.e., $p = 0$, we get our recurrence relation.

$$a_k = \frac{(k-1-n)}{k(k+m)}a_{k-1} \quad (6.93)$$

$k = 1, 2, 3, 4, 5, \ldots$

Now, take the integer values of k,

$$k = 1, \quad a_1 = -\frac{n}{(1+m)}a_0,$$

$$k = 2, \quad a_2 = \frac{(n-1)}{2(2+m)}.na_0,$$

$$k = 3, \quad a_3 = -\frac{(n-2)}{3(3+m)} \cdot \frac{(n-1)}{2(2+m)}.na_0,$$

$$k = 4, \quad a_4 = \frac{(n-3)}{4(4+m)} \cdot \frac{(n-2)}{3(3+m)} \cdot \frac{(n-1)}{2(2+m)}.na_0,$$

$$k = 5, \quad a_5 = -\frac{(n-4)}{5(5+m)} \cdot \frac{(n-3)}{4(4+m)} \cdot \frac{(n-2)}{3(3+m)} \cdot \frac{(n-1)}{2(2+m)}.na_0,$$

and so on. The explicit form of (6.80) can be written as,

$$y = a_0 + a_1 x + a_2 x^2 + a_3 x^3 + a_4 x^4 + a_5 x^5 + a_6 x^6 + a_7 x^7 + \ldots$$
$$(6.94)$$

6.7 Associated Laguerre Polynomials

Putting the value of $a_1, a_2, a_3, \ldots$ in (6.94), we get

$$y = a_0(1 - \frac{n}{(1+m)}x + \frac{(n-1)}{2(2+m)}.n.x^2 - \frac{(n-2)}{3(3+m)}.\frac{(n-1)}{2(2+m)}.n.x^3 +$$

$$+ \frac{(n-3)}{4(4+m)}.\frac{(n-2)}{3(3+m)}.\frac{(n-1)}{2(2+m)}.n.x^4 - \ldots). \qquad (6.95)$$

Now, choose the value of $a_0 = \dfrac{(m+n)!}{m!n!}$

$$y = \frac{(m+n)!}{m!n!}(1 - \frac{n}{(1+m)}x + \frac{(n-1)}{2(2+m)}.n.x^2$$

$$- \frac{(n-2)}{3(3+m)}.\frac{(n-1)}{2(2+m)}.n.x^3 +$$

$$+ \frac{(n-3)}{4(4+m)}.\frac{(n-2)}{3(3+m)}.\frac{(n-1)}{2(2+m)}.n.x^4 - \ldots) \qquad (6.96)$$

By running the indices $n = 0, 1, 2, 3, \ldots$ we get associated Laguerre polynomials

$$L_0^m(x) = 1$$

$$L_1^m(x) = (1 + m - x),$$

$$L_2^m(x) = \frac{1}{2!}\left((1+m)(2+m) - 2(2+m)x + x^2\right)$$

$$L_3^m(x) = \frac{1}{3!}((1+m)(2+m)(3+m)$$
$$- 3(2+m)(3+m)x + 3(3+m)x^2 - x^3).$$

The general expression or n^{th} term of Laguerre polynomial is given by

$$L_n^m(x) = \sum_{k=0}^{n}(-1)^k \frac{(m+n)!}{k!(n-k)!(k+m)!}x^k. \qquad (6.97)$$

We can generates the associated Laguerre polynomial using the above expression by terminating a series for integral n.

6.7.1 Properties of Associated Laguerre Polynomials

The properties of associated Laguerre polynomials basically pursue the properties of Laguerre polynomial.

Rodrigues Formula

Rodrigues formula for associated Laguerre polynomial is defined as

$$L_n^m(x) = \frac{e^x x^{-m}}{n!} \frac{d^n}{dx^n}(x^{n+m} e^x). \qquad (6.98)$$

Orthonormality

The orthogonality of associated Laguerre function is defined by a multiplicative weight factor $u = x^m e^{-x}$, i.e.,

$$\int_0^{+\infty} L_p^m(x) L_q^m(x) x^m e^{-x} dx = \left\{ \begin{array}{ll} 0 & If,\ p \neq q \\ \frac{(m+n)!}{n!} & If,\ p = q \end{array} \right\}. \qquad (6.99)$$

Generating Function

The generating function of associated Laguerre function $L_n^m(x)$ is defined as

$$g(x, t) = \frac{e^{-xt/(1-t)}}{(1-t)^{m+1}} = \sum_{n=0}^{\infty} L_n^m(x) t^n \qquad (6.100)$$

6.8 Exercises

1. Prove (6.98) using the Leibniz theorem.
2. Generates first four associated Laguerre polynomials using (6.100)

3. Prove the relation given in (6.78).
4. Find the value of N_{nl} given in (6.76) by normalizing the $R_{nl}(r)$. Hint: use the expression $\int_0^\infty r^2 R_{nl}^2(r) dr = 1$.
5. Show that $xL_n''(x) + (1-x)L_n'(x) + nL_n(x) = 0$.

Bibliography

[1] Bateman, H., Partial Differential Equations of Mathmatical Physics, Cambridge University Press, London (1959).

[2] Bateman, H., The Mathmatical Analysis of Electrical and Optical Wave-Motion, Dover Publications, Inc., New York(1955).

[3] Bowman, F., Intoduction to Bessel Functions, Dover Publications, Inc., New York(1958).

[4] Copson, E. T., An Introduction to the theory of Functions of a Complex Variable, Oxford University Press, London(1935).

[5] Erdelyi, A., Asymptotic Methods in Analysis, Interscience Publishers, Inc., New York (1958).

[6] Hochstadt, H., Special Functions of Mathmatical Physics, Holt, Rinehart and Winston, Inc., New York(1961).

[7] Jannke, E. and F. Emde, Tables of Higher Functions, sixth edition, by F. Losch, McGraw-Hill Book Co., New York(1960).

[8] Jeffreys, H., Asymptotic Approximations, Oxford University Press, London(1962).

[9] Jeffreys, H. and B.S. Jeffreys, Methods of Mathematical Physics, Cambridge University Press, London (1956).

[10] Klein, F., Vorlesungen uber die Hypergeometrische Funktion, Spinger- Verlag. Berlin (1933).

[11] Luke, Y. L., Integrals of Bessel Functions McGraw-Hill Book Co., New York(1962).

[12] Morse, P.M. and H. Feshbach, Methods of Theoretical Physics, McGraw-Hill Book Co., New York(1953).

[13] Mathematical Methods for Physics and Engineering, K. F. Riley, M. P. Hobson and S. J. Bence, third ed., Cambridge University Press, (2006).

[14] Quantum Mechanics, Eugen Merzbacher, second ed., Wiely Publication.

[15] Mathematical Methods For Physicsts, George B. Arfken and Hans J. Weber, sixth ed., Elsevier Academic Press Publications, (2005).

[16] Quantum Mechanics concepts and application, Nouredine Zettili, second ed., Wiely Publication.

[17] Quantum Mechanics Theory and Applications, Ajoy Ghatak and S. Lokanathan, Springer Science+Business media, B.V., Kluwer Academic Publisher.

[18] Laguerre Polynomials Chapter 7-Module 2, Sandip Banergee, e-pathsala.

[19] Optics, Ajoy Ghatak, McGraw-Hill Publisher.

Index

A

Application to Probability Theory, 23

Associated Laguerre Polynomials, 97, 98, 101, 102

Associated Legendre Functions, 44, 45

Asymptotic Representation of the Gamma Function, 10

Asymptotic Representation of Probability Integral, 17

B

Bessel Functions, 49, 105

D

Definite Integrals, 7, 11, 37

F

Fresnel Integrals, 20, 21, 22, 28

G

Gamma Function, 1, 3, 6, 11, 52

Gamma Function and Some Relations, 3

Generating Function, 54, 69, 73, 85, 102

H

Heat Conduction, 16, 25

Hermite Functions, 65, 68

Hydrogen atom, 94, 96

Hypergeometric Equation, 32, 35

I

Integral Representations, 29, 37, 38, 46

L

Laguerre Functions, 81

Legendre Functions, 35, 37, 39, 44

Logarithmic Derivative, 6

O

Optical Fiber, 60

Orthonormality, 59, 76, 92, 102

P

Probability Integral, 15, 16, 17, 18, 29

R

Recurrence Relations, 42, 57, 70, 87

Rodrigues Formula, 73, 74, 91, 102

S

Simple Harmonic Oscillator, 76

T

Theory of Vibrations, 23, 26

W

Workskian, 40

About the Authors

Bipin Singh Koranga is a Graduate from Kumaun University, Nainital. He has been with the Theoretical Physics Group, IIT Bombay since 2006 and received the Ph.D. degree in Physics (Neutrino Masses and Mixings) from the Indian Institute of Technology Bombay in 2007. He has been teaching basic courses in Physics and Mathematical Physics at the graduate level for the last 12 years. His research interests include the origin of universe, Physics beyond the standard model, theoretical nuclear Physics, quantum mechanical neutrino oscillation and some topics related to astronomy. He has published over 42 scientific papers in various International Journals. His present research interests include the neutrino mass models and related phenomenology.

Sanjay Kumar Padaliya is presently Head, Department of Mathematics, S.G.R.R. (P.G) College, Dehradun. He received his Ph.D. degree in Mathematics (Fixed Point Theory) from Kumaun University, Nainital. He has been teaching basic courses in Mathematics at graduate and postgraduate level for the last 20 years. His present research interests include Fixed Point Theory and Fuzzy Analysis. He has published over 25 scientific papers in various International Journals of repute and has presented his works at National and International conferences. His latest book, "An Introduction to Tensor Analysis" is published by River Publishers. Dr. Padaliya is also a life member of Indian Mathematical Society, Ramanujan Mathematical Society and International Academy of Physical Sciences.

Vivek Kumar Nautiyal is graduate from Lucknow University, Lucknow. He did his M.Tech in Applied Optics from Indian Institute of Technology Delhi in 2014. He has remained up-to-date with Nuclear and Particle physics since 2014 and earned his Ph.D degree in physics (Neutrinoless Double Beta Decay) from Babasaheb Bhimrao Ambedkar University (A Central University), Lucknow. He has more than three years of teaching experience at graduate and postgraduate level. His research interests include Nuclear and Particle Physics, Neutrino Physics, Physics beyond the standard model and Optics.